Farming, Friends, & Fried Bologna Sandwiches

Advance praise for

Farming, Friends, & Fried Bologna Sandwiches

Like grandmother's pie sitting in the windowsill, this book will entice you with its stories and love of Southern food.

—Nathalie Dupree, author of *Mastering the Art of Southern Cooking*

Renea Winchester won the hearts of readers with her glorious book, *In the Garden with Billy*. Now she's penned a marvelous sequel plumb full of authentic true-as-gospel Southern country folks, their lives, foods, recipes, and unique way of life. The chapter on goobers is worth its weight in gold. As we say in the South, "It's a dandy."

—Joseph E. Dabney, author of *Smokehouse Ham, Spoon Bread and Scuppernong Wine*, winner of the James Beard Cookbook of the Year Award

If you crave a real, authentic trip down South, complete with mouthwatering Southern food, Renea Winchester's latest, *Farming, Friends, and Fried Bologna Sandwiches* will have you salivating for a rocking chair on a big front porch. But it's the people and the "true stories, stretched truths and downright lies" that keep the pages turning. The detailed and graphic pictures she draws of the experiences she has shared with her farming friends and life in the South will satisfy the soul. This visual gem just makes you hungry for more of Renea Winchester's story-telling magic.

—Beth Albright, actress and award-winning author of the Sassy Belle series

My ninety-six-year-old grandmother was a firm believer that there is no substitute for the benefits of getting your hands dirty. She taught me that hard work is good for the soul; a garden is a balm for the heart. Gardening is about simplicity: a human being cradling nature in their hands and nurturing it. Renea and Billy seep that kind of love from their pores, they dream it in their sleep. Their message is a powerful one—that caring for a goat, a garden, the land, or the community is all equally important to the Earth.

—Shawna Coronado, community activist, gardener, difference maker

Georgia farmer Billy Albertson, whom readers have come to know and love through Renea Winchester's writing, is no chef. He uses his oven for storage. But his kitchen is a hallowed place, frequented by friends who may stop by to deliver a little supper or to enjoy a midday tomato sandwich, but also always to chew the fat. In Billy's kitchen, "dope" is still a cold cola and the only PhD that matters is a post-hole digger.

Farming, Friends, and Fried Bologna Sandwiches, Winchester's follow-up to *In the Garden with Billy,* is part gardening guidebook, part laugh-out-loud memoir, and part Southern-cooking primer. She shows us, over the fellowship of a fried baloney sammich cooked in a cast-iron skillet, the nearly forgotten connections between farmer and food supply, shared meals and shared experiences, cooking and our common culture. Southern recipes are valued the world over because they are simple and real, because they have been tested and improved with time, and because they always come with a story. A satisfying Southern meal, like Billy himself, is unapologetically uncomplicated—yet somehow speaks to us at a depth we may not expect.

—Deborah Geering, food writer, local-food advocate, relocated Yankee

Made from scratch with plenty of love and attention, Winchester serves us a slice of the Southern good stuff in this collection of wisdom pulled straight from Billy's garden and kitchen. With sentimental dirt-beneath-your-nails tales about cooking everything from potlikker and fried bologna sandwiches to black-eyed peas and hog jowls, there is nothing fancy here. Just a real down-home connection to the food we eat, the farmers who grow it, and the bonds we build around the supper table. No doubt, readers will be shouting a heartfelt Amen and asking for seconds.

—Julie Cantrell, *New York Times* and *USA Today* bestselling author of *Into the Free* and *When Mountains Move*

In the winning voice of a friendly farm neighbor, Renea Winchester shares her down-home wisdom on all aspects of the Southern garden and table, from seeds to fried bologna. Her tips are as warm and practical as Georgia sunshine, her applications to life, friendship, and love, universal. A great gift book for a budding gardener, newlywed cook, or to read on the porch with a glass of sweet tea.

—Janis Owens, author of *American Ghost: A Novel* and *The Cracker Kitchen*

Work hard, rest hard, love hard. Spend time with friends. Eat homegrown tomatoes and fried baloney sandwiches. A flawless recipe for a good life. Renea Winchester delivers a simple but powerful *how* to employ this timeless wisdom through her muse, the inimitable Billy Albertson.

—Jill Conner Browne, #1 *New York Times* bestselling author of *The Sweet Potato Queen*

Written with charm and grace, *Farming, Friends, and Fried Bologna Sandwiches* is both a tribute to a vanishing way of life and an informative resource about traditional farming that belongs on every family's bookshelf. In this sequel to *In the Garden with Billy,* Winchester continues her quest to preserve the knowledge of lifelong Georgia farmer, Billy Albertson. For those who have spent time on a traditional farm, these unvarnished stories of working alongside Billy will bring back memories of a challenging yet deeply rewarding way of life, while those who are unfamiliar with farm life will find this book to be an eye-opener.

—Amy Hill Hearth, *New York Times* bestselling author of *Having Our Say: The Delany Sisters' First 100 Years* and the critically-acclaimed novel, *Miss Dreamsville and the Collier County Women's Literary Society*

Before you begin to read *Farming, Friends, and Fried Bologna Sandwiches,* you should take a moment to prepare yourself for a life-changing experience. As was the case with her previous book, *In the Garden with Billy,* Rene Winchester touches upon those facets of human nature that represent the best in all of us: love, faith, duty, purpose, family, and tradition. *Farming, Friends, and Fried Bologna Sandwiches* makes me want to be better than I am. As I read the touching narratives of this uncomplicated country man, I found that I wanted nothing so much as to go home once again to a place I lost long ago, to work an honest day on a small piece of ground, and to sit in the shade in the evening, watching the sun drop down behind the hills.

—Raymond Atkins, author of *Camp Redemption: A Novel*

FARMING, Friends, AND FRIED BOLOGNA SANDWICHES

Renea Winchester

MERCER UNIVERSITY PRESS MACON, GEORGIA
35 YEARS OF PUBLISHING EXCELLENCE

MUP/P494

Published by Mercer University Press, Macon, Georgia 31207

1400 Coleman Avenue
Macon, Georgia 31207

9 8 7 6 5 4 3 2 1

Books published by Mercer University Press are printed on acid-free paper that meets the requirements of the American National Standard for Information Sciences—Permanence of Paper for Printed Library Materials.

Photographs on pages 34, 148, 220, 238, 290
courtesy of Tracy Hoexter Photography
Photographs on pages 140, 188, 264, 274
courtesy of Kelle Mac Photography
Photographs on pages 170, 210, 282 courtesy of Ana Raquel
Photographs on pages 84, 110, 140, 160, 228
courtesy of Renea Winchester

Library of Congress Cataloging-in-Publication Data

Winchester, Renea
Farming, friends & fried bologna sandwiches / Renea Winchester
pages cm
Includes index
ISBN 978-0-88146-504-4 ISBN 0-88146-504-6 (pbk: alk. paper)
1. Farm life–Georgia. 2.Cooking--Southern States. 3. Albertson, Billy.
4. Winchester, Renea. 5. Albertson, Billy. I. Title

S521.5.G4W56 2014
630.975–dc23
2014023717

Contents

With Sincere Thanks

I begin with a heartfelt expression of appreciation to those who read *In the Garden with Billy: Lessons about Life, Love & Tomatoes* and encouraged me to tell them more. This is for you.

To Wanda and Nicki of the Southern Independent Booksellers Alliance and the Indie Booksellers who breathed life into my work, who held events in your stores, and who sold copies of my book, thank you. To the book clubs, garden clubs, community clubs, and others who invited Billy and me into their lives and then shared my words with others, I am forever grateful. For my Facebook friends who encouraged me during the early-morning and late-night hours, especially the Aussie and UK friends, I am honored and humbled that you enjoyed our story. To the Pulpwood Queen, Kathy Patrick; the Sweet Potato Queen, Jill Conner Brown; and Queens the world over, power to the written (and printed) word. Stories will never die.

Many of you know that libraries hold a special place in my heart. They are the gateway to imaginations, a safe haven for children after school, and, many times, the only access adults and students have to the Internet. So let me pause for a moment to hug the public library staff who work tirelessly, fueled by passion and caffeine. Thank you. Thank you from all of us for everything you do, especially those who allowed me to speak at your facility.

Thanks to my fellow authors who inspired and pushed me along, whom there are too many to name. I reached out, begged for your help, and you responded. I am in your debt. To members of my critique group who provided honest feedback, your encouragement keeps me writing. To Richard and Grant, I listened. To Ana Raquel who translated Frank's words and captured her father's visit from Puerto Rico on film, thank you. Ann Hite, you will never know how much our conversations inspired me. Without our chats, the manuscript would remain unfinished. A big ol' thank-you to Raymond Atkins, who gave his valuable time to me, a fledgling. To Laurie, who discussed books over cupcakes, and Beverly, who sprinkled a bit of magic. Thank you to my early readers, Amy Hill Hearth and Julie Cantrell. With deepest thanks to Karen Spears Zacharias, who made the introduction; and to Marc Jolley, who said yes. Bushels of love poured out on the entire Mercer University Press team. I am proud to call Mercer University Press home.

Squeezes to the sisters I have met at Billy's. To Billy's daughters, Janet and Denise, thank you for sharing your daddy with me. And thanks to Kelle and Tracy, who use their photography gift to capture images of life on the farm. I'll see you in the field!

Extra hugs to Lamar, Austin, "Neighbor Joe," and the late Andrew Wordes, who help Billy any time his "ox gets in the ditch." The farm would not function without you and the assitance of others who pitch in.

For Billy's customers who buy tomatoes, figs, and anything else he happens to grow on the little strip of land he calls home, thank you. For the many helpers who were so inspired after reading about Billy that they donned work gloves and lent a hand, thank you. To Officer Bracken who watches over Billy, his daughters and I are grateful. And for those who understood the importance of relationships and reached out to their own Billy Albertson, thank you for giving your time to someone in need.

Saving the best for last, to my beloved husband Dennis, who is my rock, my fortress, my all. You believed, you shared me with the readers, and you have patiently waited. You know my heart, and for that I am thankful. I love you. To Jamie, the beginning of it all: star, heart, smiley face.

A Word about the Recipes

This book is a compilation of gardening experiences and conversations exchanged while working beside farmer Billy Albertson. It is a memoir and a biography. It is a how-to-garden book and a cookbook. Most of all, it is a piece of my heart given to you, the reader, in appreciation for inviting me into your life. I hope you enjoy this story enough to share it with others.

Billy's kitchen isn't exactly functional. I am certain that his wall oven could bake a delicious cake and chicken pot pie, but currently it serves as storage. For that reason, I carry food from my home to his, as do many other garden helpers and lifelong friends. When Billy and I traveled together after the release of the first book, we met readers who lavished Billy with affection and spoiled us with delicious homemade food. One particular club member spent hours decorating a chicken-shaped cake. We met many folk who loved bologna sandwiches, and others who had never eaten one.

"Folk just don't know what they're missing," Billy said, referring to a fried bologna sandwich. I agreed, which is why I determined it was my duty, actually my privilege, to share our favorite recipes with you.

How can one measure a Southern serving size? It isn't possible. For example, Billy and I enjoy two pieces of bologna on our sandwiches, while others enjoy them made with just one slice. In Billy's kitchen, it is perfectly acceptable to have a heaping helping and then return for seconds. Or, as my Grandpa used to say, "Eat until your tongue smacks your brains out." For that reason, it is difficult to gauge exact serving sizes for the recipes found here; they are approximate.

A note about spelling: I had lengthy discussions with the editors at Mercer University Press about the proper spelling of bologna versus baloney. Depending on your preferred dictionary, spellings and definitions vary. Webster cites baloney as "foolish words or ideas," while the American Heritage Dictionary (1978), defines baloney as "(1) *Informal.* Variant of bologna; (2) *Slang.* Nonsense." Dear ones, it appears that there is quite the controversy over this little slice of lunch meat.

Many readers grew up hearing the Oscar Mayer jingle and have "bologna" deeply rooted in their brains. Realizing that spellings vary depending on geography, I took to Facebook for an "official poll," which proved my theory that most Northerners prefer "bologna," while here in the South, we are loyal to "baloney." And then there are my Appalachian folk, who pronounce it 'loney. Ultimately, I wanted my words to ring true to the people who have opened their lunchboxes and consumed this delicacy, to mothers who fried bologna for their kids, to those who have worked hard all their lives and know that fried baloney sandwiches are not only fuel for the body but also sustenance that triggers memories from the past.

Perhaps Henry Lee Smith, Jr., says it best in the *American Heritage Dictionary* in the special article titled "Dialects of English": "It is obvious to all of us that people differ in the way they speak the same language. No regional or social dialect can be singled out as the only correct form of a language...."

What I think Henry Lee Smith, Jr., means is that disagreeing over the spelling, or pronunciation, seems trivial when there are fried bologna sandwiches waiting to be enjoyed.

Finally, I spend most of my time with Billy working his little strip of land, not cooking in his kitchen. A majority of the stories happen there. The recipes found in this book are, as we say in the South, just gravy.

FARMING, Friends, AND FRIED BOLOGNA SANDWICHES

1

The Tradition of Bologna Sandwiches

There is no shame in enjoying a fried bologna sandwich. Some foods trigger memories. Whether we're smelling a peach or trying sushi for the first time, food binds our taste with our experiences. Food memories, good or bad, linger in our adult lives. I bet you can still remember the first time you tasted a gooey campfire s'more dripping with melted chocolate and marshmallow fluff: the feel of a rough graham cracker as it touched your fingers; the anticipation as you pressed the crackers together, blending chocolate with puffy white goo. Your tongue traced the edge of the cracker. You wanted to savor each bite, but then your best friend said, "Bet you can't cram the whole thing in your mouth."

So you did.

Billy Albertson loves bologna sandwiches. It does not matter how Oscar Mayer spells B-O-L-O-G-N-A, for Billy it's "baloney." In his day, fried "baloney" sandwiches were a delicacy. They still are today.

Stereotypes label Southerners with an advanced level of outdoor expertise. Southerners can kill a buck, spit tobacco juice through gapped teeth, wrestle alligators, and survive in the woods while wearing only a coonskin cap and carrying a pocketknife. Truth is, few Mason-Dixon Line residents enter the wilds of nature intent on snaring an animal with which to feed their family these days. But we sure do enjoy a good fried bologna sandwich.

When I was growing up, Mother rarely scored the bologna. My brother and I waited on tiptoe for the precise moment when the meat formed the shape of a cap one might wear on his or her head. We held out our hands as Mother speared a slice with a fork, and then we nibbled half-moon bites around the slice. While we devoured the first piece, she fried another for our sandwiches. You can't buy memories this delicious.

It is this memory, coupled with the joy of sharing a sandwich with loved ones, that sears the desire for this particular food deep into my heart.

When mountain folk drive into the woods, they go prepared. They load the pickup truck with tents, stack a couple cords of firewood in the back, and fill a cooler with adult beverages. They park at the Piggly Wiggly and stock up on traditional must-have snacks that most people shun: Spam, potted meat, bologna, Beanee Weenee's, and tiny fish in a can that some people call "little kippers." Later, as the campfire embers begin to glow, the men open cans of pork-n-beans and arrange them in a circle around the fire. They seal slices of bologna inside aluminum foil and position the shiny ovens near the heat. Then it is time to kick back, cross one leg over the other, and pop the top of a Budweiser. True stories, stretched truths, and downright lies swirl with the smoke as hardworking folk relax in lawn chairs and wait for steam to rise from the open cans. During this wilderness experience, men will not shave, children will not bathe, and women will endure it all, mindful that these moments forever bind together family and friends.

City folk label outdoor experiences as quality time. Mountain folk call it life.

&

My warm and fuzzy memories of bologna transport me back to the 1970s, when Mr. Conner delivered Sunbeam bread fresh each morning, and big-box stores hadn't yet encroached into the mountains of western North Carolina. Leaning across the counter of Winchester's Grocery, a country store my grandpa, Frank Winchester, operated, I would create a mental list of the cigarettes he needed to restock the shelf: Marlboro, Kool Super Long Menthol, Pall Mall, Camel, Vantage. Eager to help, I returned from the storeroom with cartons stacked beneath my chin. Dodging Grandpa's unfiltered Camel cigarette that was burning a track in the counter, I ripped open the cartons and restocked the shelf. I was ten years old.

Even though Grandpa is in heaven, and Winchester's Grocery is closed, I can still remember my time spent at the store just like it was yesterday. Often we stopped at the store because Mother had a hankering for a "dope in a bottle." Down South, some folk also called this beverage a "Co-cola." Mother's midday pick-me-up was a Coca-Cola with a bag of Lance peanuts poured inside—a perfect combination of salty and sweet in every sip.

Tightly packed in a red, white, and blue Pepsi-Cola cooler, ice-cold bottles of Coke, Nehi Grape, Mountain Dew, Pepsi, and Dr. Pepper waited for thirsty patrons. Back then, store space wasn't proprietary. Even bottles lived in harmony.

Customers trickled into the store, paying on their accounts while simultaneously charging more groceries to their tabs. Weary men and women who were dirty from factory work, the ink on their checks barely dry, picked up a few items on their way home. In those days, there was not a Walmart, Target, or a super-sized grocery store in my hometown of Bryson City, North Carolina. Thankfully, there still isn't.

Conveniently located on a busy road near the Alarka community, Winchester's Grocery offered a variety of canned goods, dairy products, animal food, tobacco, and an assortment of medical supplies available even in the wee hours of the morning when young children tend to spike high temperatures. Many times, Frank Winchester would slip his feet into spit-shined shoes and drive to the store in the middle of the night. Worried mothers ignored the clock when they needed to make a poultice and were out of Save the Baby medicine.

Ripping the cartons open one day, I fed cigarettes into a dispenser. I fought the desire to alphabetize the brands. I had tried organizing the cigarettes before, but Grandpa assured me that Camel and Marlboro belonged together within arm's reach.

"Some customers want to grab a pack of smokes and be on their way," he said, understanding his customers' shopping habits far better than a modern-day computer ever could.

With no time to return the thick roll of bologna to the cooler, Grandpa wrapped it with wax paper and secured it with one of the bright blue rubber bands Granny stockpiled in the catchall drawer. Desiring to be part of important grown-up activities, I reached for the shiny red roll of bologna and asked, "Want me to put this back in the cooler?"

Grandpa shook his head and nodded toward the line of people waiting. "Naw. Somebody will just carry it right back up here. Best leave it where it is."

Cold cuts weren't sold in the shrink-wrapped packages one finds in today's grocery store, but rather by the hunk, sliced with a blade that resembled a bayonet. After twenty years of operating Winchester's Grocery, Grandpa inherently knew how much his customers wanted. When someone asked for a hunk of bologna, he grabbed the knife, positioned it on the red protective cover, and then looked at his customer, who either acknowledged that the size was adequate or indicated more or less by holding up fingers for him to reposition the blade. In one swift movement, Grandpa sliced. Wax paper crinkled as he plunked the bologna on top of it and asked, "Need some cheese to go with this?"

Bologna sandwiches still typify my mountain people: folk who pack strong coffee into a silver thermos, who raise families on minimum wage and store credit, who harvest tobacco during the summer and hang it from barn rafters to cure. These people relish the smell, the taste, and the meaning of home.

&

No matter how old you are, you can still get homesick. Moving from a small town into a busy metropolis is a daunting venture. After growing up surrounded by people who knew me, I struggled with the aloneness that comes from living outside of your raising. I quickly learned that city folk are transient. They move from hither to yon mostly based on the whims of corporate managers and company presidents. They buy enormous homes and invest large amounts of time and energy by erecting walls around their hearts. I had never met a stranger until I moved into the big city. My attempts to befriend them left me feeling inadequate. Luckily, there are no strangers on Billy Albertson's farm. On his little strip of land, you can be yourself.

I met Billy several years ago after my daughter, Jamie, noticed a sign with the words BABY GOATS. Since that time, I have had several adventures with him. I have joyfully cried when baby goats were born and wept when they died. With Billy, I have learned and laughed and finally penned a book titled *In the Garden with Billy: Lessons about Life, Love & Tomatoes*. The book brought many new friends to his door, and, as the cliché goes, the rest is history.

ꝏ

Shortly after the release of my first book about Billy Albertson, Billy endured radiation and hormone shots for an aggressive form of, using his words, "prostrate" cancer. The short-term side effect was lethargy. Long-term side effects included intolerance of heat, sporadic yet severe hot flashes, insomnia, and the inability to eat many of the foods he had enjoyed for years.

Though incredibly unhealthy, and not intended for daily consumption, a fried bologna sandwich warms the heart, especially when shared with friends. So when Billy Albertson first plunked a cast-iron skillet on top of his stove and said, "How 'bout a fried baloney sammich for lunch?" I knew I had found a little slice of home in the heart of Atlanta. I soon discovered that fried bologna sandwiches are a delicacy in Billy Albertson's home. While the sizzle of nitrate-laced fillers dancing in a cast-iron skillet causes vegans and health-conscious foodies to cringe, Billy and I understand what they are missing. This knowledge brings a smile and an open invitation for visitors to gather around his Formica table.

For a man who grew up eating turnips, collards, dried beans, and deep-fried everything, making culinary lifestyle changes is difficult. Billy, like any other eighty-year-old widower, has learned to make do and make dinner. Fortunately, many people read my first book and fell in love with him. They wanted to make certain he lived a good, long life. Visitors arrived bearing gifts of food. They carried nutritious, healthy foods such as salads and fruit smoothies; food that Billy ate with gusto; food that eventually made him very ill.

Not to divulge too much information, but a condensed, compressed, juiced-up healthy diet wreaked havoc on the man and almost landed him in the hospital. After speaking with Billy's daughters about his digestive condition and relaying details about his daily food intake, it became apparent that Billy, who would never hurt the feelings of another, should kindly explain to his new friends that a regular diet of fried bologna sandwiches was just what the doctor ordered.

Honest, it is.

When Billy was growing up, food was never wasted. He ate everything his mother placed on the plate. After years of this "waste not" lifestyle, educating him about food was a process that involved physicians, family, and friends. Months of radiation treatments left him with what he calls a "tender gut" that needed to heal. Concerned friends and family knew that his condition called for a delicate balance that protected Billy's tummy while encouraging those who adored him to help. Billy truly enjoyed the gifts and savored every mouthwatering morsel. Unfortunately, even years after the treatments, he still cannot tolerate greens (meaning salads, collards, kale, and mustard greens). When hunger pangs hit, Billy often heads for the pantry in search of a box of Saltine crackers and a can of Vienna Sausages, which he pronounces "Vi-enny sausages."

Lawd have mercy. Just the thought of a Vienna Sausage makes my tummy feel icky.

Stocking the kitchen with nutritious foods was a challenge. Fortunately, Billy has an agreement with one of his precious neighbors. She cooks more than she can consume. Billy loves to eat. This is a perfect fit. Another friend, Mrs. Davis, spends hours preparing lunch for Billy. He also has a customer who delivers what is the ideal meal: homemade chicken pot pie. Even with this arrangement, there was a time when Billy's cooks experienced their own periods of ill health. Feeding Farmer Billy became problematic.

Of course, his daughters provide tummy-pleasing foods such as yogurt and almond milk. They also keep a close eye on him. His firstborn, Janet, spends a majority of the hot summer months standing in a sweltering kitchen canning beans and freezing vegetables so Billy can enjoy a quick meal of okra or pinkeye purple hull peas during the winter. Denise, whom Billy calls "Sweetheart," monitors the kitchen hardware, especially the working condition of the Crock-Pot, which Billy uses but neglects to attend. Neglect is a harsh word. What really happens is that he gets too busy in the garden or nods off during a midday nap, waking only when the acrid smoke from a burned dinner sounds the alarm.

Realistically, his daughters cannot be there all the time. Besides, those Vienny Sausages are tempting and so simple to prepare.

Two of Billy's favorite foods are rice and the deliciously Southern, piping hot pone of cornbread. He grows the corn himself and knows the person who grinds the meal. The issue is always time and maintaining a watchful eye on what is cooking. During the summer, produce-stand customers lure him away from domestic duties. According to Billy, he does not have time to fool with watching the oven. This is why visitors often find the front door open and a box fan perched on the threshold, belching smoke out of the kitchen.

Monitoring Billy meant incorporating a host of volunteers and a hint of trickery. By diplomatic vote, we appointed fellow farmhand and health nut extraordinaire, Kelle McEntegart, to serve as the Quality Control Refrigerator Specialist. Her job is making executive decisions regarding the contents of the Albertson refrigerator. Thus far, Billy seems comfortable with this arrangement. On occasion, he even says, "The refrigerator is a mess. Looks like I need to call Kelle to come over here and clean it out." She arrives bearing a loaf of freshly baked organic bread. While searching for real butter to slather upon the still-warm slices, she rummages through the contents of Billy's refrigerator and tosses unacceptable items into the trash.

Predictably, she discovers a mega-sized tub of margarine. With an angry frown and a joking-yet-serious wag of the finger, she gives him the business.

"Where did that disgusting margarine come from?" she demands. "You know it isn't good for you."

Billy smiles, nods, and enjoys a hunk of real butter on whole wheat bread, knowing that a container of low fat cottage cheese safely hides our bologna stash.

Don't tell Kelle. We all have our secrets.

ᘓ

Shortly after his last hormone treatment, Billy proclaimed that it was time to "break the faller ground and plant the spring crop." We had both struggled that year. I believe he pushed himself too hard, too soon, as evidenced by episodes he called "weak spells" that came without warning. We would be working in the garden, and then he'd suddenly bend over, hands on knees, gasping for breath.

"I just went weak all the sudden," he explained as I rushed to place a chair beside him so he could rest. I patiently waited while he said, "I'm just no good to anyone no more." I offered him encouragement and a glass of cool water.

What more could I do?

I also worried about air quality and Billy's struggles to breathe during the heat of the day. Smog alerts are common in the area. Although medical tests revealed no lung anomalies, his coughing and wheezing spells continue to cause me great concern.

Still, as painful as Billy's weak spells are to witness, resting with him until the moments pass is excruciating. Eager to get out of the sun that particular day, I seated Billy beneath the pecan tree. He patted the lawn chair and said, "It's break time. We've worked hard this morning." Even though I was capable of completing the necessary chores without him, I obeyed. My desire was to finish the chores quickly while Billy rested. But Billy didn't want to finish fast. He wanted company until the weak spell passed. Besides, Billy's health was more important than gardening.

The doctor expressed no concern about the weak spells and ordered Billy to take it easy. This type of medical advice was about as effective as those who dictate that I should limit my chocolate intake to once a week. Billy's idea of taking it easy involved the strategic placement of lawn chairs throughout the garden. He worked until he felt dizzy, then collapsed into a chair.

It is certainly difficult to keep an outside man inside the house. I respect his pioneer lifestyle. With a coordinated plan of helpers and strategically scheduled visitors, we have somehow managed to work the garden, rest the farmer, and operate the produce stand that has made him a local celebrity.

Billy Albertson touches many lives. Those who care about him want to be certain that he maintains his lifestyle as long as possible. With the cancer treatments behind him, Billy's stamina has partially returned.

Today, the lawn chairs that once adorned the vegetable rows and served as a soft place to land are arranged in a semicircle, waiting for guests who visit. Truly, it appears that Billy has won the battle with cancer. When we bow our heads and give thanks for fried bologna sandwiches, we also open our hearts with grateful thanks for the people God places in our paths.

Billy often says, "I'm a lucky man to have so many friends."

I disagree. He deserves the gift of friendship.

Curing the Cast-iron Skillet

Most of the recipes found in this book are prepared using a cast-iron skillet. While nonstick pans may be the most modern way to prepare meals, every Southern home needs a cast-iron skillet.

The Curing Process

- Preheat oven to 350 degrees.
- After purchasing a skillet, wash and dry the pan. Pour enough vegetable oil to cover the bottom of the pan and rub onto the sides.
- Place skillet in heated oven for one hour. The pan may smoke, but do not be alarmed. Several times during the process, remove the pan and make certain that the bottom and sides are still oiled. If not, carefully add more. After an hour, turn off oven and allow skillet to cool. Use a towel to remove any excess oil.
- The skillet is now ready to use. Repeat this process if food begins to stick during cooking.

Care and Maintenance

- Keep your skillet dry. A damp cast-iron skillet will rust. Never place the skillet in the dishwasher or leave it submerged in water for an extended period.
- Clean the surface using a damp cloth. Wipe dry with a paper towel, then place the skillet upside down in a warm oven for five minutes. Store skillet in oven until next use.

Fried Bologna (or "Baloney") Sandwiches

Billy buys the cheapest bologna in the grocery store—usually a brand that requires him to break a safety seal and peel off a red wrapper prior to frying. However, money is no option when purchasing mayonnaise; it is Duke's or nothing.

This quasi-healthy recipe uses quality bologna from the deli that is sliced thick for sandwiches. Hebrew National makes an excellent kosher product. These sandwiches are extra yummy when adorned with thinly sliced cheddar cheese and partnered with Dilly Beans (recipe follows).

Ingredients

1 teaspoon Duke's Mayonnaise (no Duke's=no sandwich)
2 slices white bread
2 slices thick-cut deli-style bologna
1 teaspoon mustard (if desired)
Nonstick vegetable spray

Spread Duke's Mayo on one or both slices of bread. Set aside. Preheat a cast-iron skillet on medium-high heat. Spray skillet with non-stick vegetable spray. Lightly score each slice of bologna so it will lie flat while cooking. Place slices in the skillet, and fry each side approximately one to two minutes. The goal is to brown the bologna without burning it. Expect the meat to darken in the center. Place on white bread and enjoy immediately.

Allow me a moment to ramble about mayonnaise. Admittedly, I've lived in the South all my life. But the South seems divided into several regions of loyal mayo connoisseurs. There's the JFG clan: those who primarily bleed orange, shout "Go Vols" with 102,000 of their closest friends, and are die-hard Gator haters. Cajuns have a particular affection for Blue Plate, and the rest of the South remains loyal to Duke's. We cannot abide the sweetness of Miracle Whip, which is "salad dressing," not the creamy condiment that makes all sandwiches taste just right.

No offense, but one can spot a Yankee in the grocery store from a mile away. During a recent trip to my hometown, I stopped by Ingles grocery, where I ran into an old friend of mine. As we were chatting and blocking the aisle right beside the mayo selections, a shopper approached. Being a polite Southerner, I moved my buggy to allow her access.

"Don't bother," she said gruffly. "I'm not looking for Duke's. I married a Yankee. I have to buy Hellman's. This is what I get for not listening to my momma." She flung the jar into the buggy and, like most small-town folk, parked her buggy and shared a story. "I tell you right now, I drew the line when I caught him putting mayonnaise on pinto beans and then sprinkling them with sugar. Told him I'd stop making them if I ever seen another drop of either on my beans."

I nodded, understanding completely.

"And the cornbread," she said, spitting the words, "let me be clear. I have not, nor will I ever put sugar in my cornbread. Mercy, sometimes I wonder about my decision to marry this man. Then I caught him putting sugar on the green beans."

I'd heard of the Yank's inclination to sweetened cornbread, but sugar on the hallowed bean? That's blasphemous. "He didn't," I said while clutching my chest.

"Yes ma'am, he did," she said while turning her buggy. "I didn't tell Momma."

Dilly Beans

I first tried dilly beans at the Sawmill Hill Freewill Baptist Church in my hometown of Bryson City, North Carolina. As ladies unwrapped fried chicken, potato salad, and a bounty of other made-from-scratch dishes, Annie Mae Cooper popped open three wide-mouth jars.

"What are those?" I asked as she placed a jar on each table.

"Dilly beans," she replied.

I am certain my face revealed confusion. Perhaps I even turned up my nose a bit, which is why she pierced two beans with a silver fork, cupped her hand to collect the dripping brine, and approached.

"Just try 'em."

Crisp and filled with garlic flavor, dilly beans are delicious. They are so scrumptious that they have replaced pickles at my house. Annie Mae, thank you for sharing this recipe. It is with great pride that I pass it along to others. If you like dill pickles, you will love this recipe.

Supplies
4 to 6 pint jars, rings, and can lids

Ingredients
2 pounds unbroken green beans (washed with stems and strings removed)
4 heads fresh dill
4 cloves garlic
2 ½ cups water
2 ½ cups white vinegar
¼ cup pickling salt or kosher salt

Combine water, vinegar, and salt in a saucepan. Heat until mixture begins to boil and salt has dissolved. Set aside to cool. Place a small sprig of dill and one slice of garlic in the bottom of a glass canning jar. Tightly pack beans lengthwise inside.

The best way to pack the jar is to lay it on its side. Gather beans in your hand and slide them into the jar. Add more beans until you think the glass is full; it is probably not. Slide a knife along the edge and press hard. You'll probably find that there's room for three or four more beans. You want the jar packed tightly or the beans will float around inside when you add the brine.

Pour liquid into jar. Leave ¼ inch of space at the top. Wipe the jar opening with a clean cloth to remove any trace amounts of moisture. Place a lid on the top, secure lid, and tighten ring just enough to seal the jar.

Sealing Instructions

Dilly beans are one of many vegetables processed in a water bath. This means that no pressure cooker is required. You simply need a large pot with enough headroom to cover the jars with one-half inch of water.

To prevent glass from breaking, pour warm water into the pot. Add jars. Make certain that the water covers the jar lids. Cover pot with lid and heat water until boiling. Boil in water bath for ten minutes.

If water boils too much (causing jars to slam together), decrease temperature. After ten minutes, carefully remove the jars. Be cautious of escaping steam that can burn the skin. Use metal tongs to remove and then place jars on a towel to cool. When the lids make that unmistakable popping sound, they have sealed.

Allow three weeks for the flavors to mesh before serving.

Chicken Pot Pie

This is one of Billy's favorite dishes. A special thanks to the women who have prepared this for him throughout the years.

Crust Ingredients
2 cups plain, all-purpose flour
1 teaspoon salt
¾ cup vegetable shortening
6 tablespoons ice cold water
NOTE: If you are pressed for time, a store-bought crust will suffice.

Pie Filling Ingredients
3 cups cooked chicken, skin removed and cut into bite-size pieces
1 can cream of chicken soup
1 cup water
1 teaspoon black pepper
½ teaspoon rosemary
½ teaspoon garlic powder
2 cans chicken broth
1 (8 oz) container fresh mushrooms (sliced)
3 large carrots, sliced
2 stalks celery, chopped
1 small onion, chopped
2 potatoes, peeled and cubed

Crust

In a large bowl, combine 2 cups flour and salt. Then, using a fork or pastry knife, cut in the vegetable shortening until mixture resembles coarse cornmeal. Add water until you can shape mixture into a ball. Divide into two sections. Reserve one for the top crust. Roll the other ball flat and press it into a 9-inch pie plate. Score the crust and place into 400-degree oven for 5-7 minutes.

(continued next page)

Filling

While lower crust cooks, prepare pie filling. In a saucepan, add water, cubed potatoes, and carrots. Boil for 5 to 7 minutes. Precooking thick vegetables shortens baking time.

Drain vegetables and place in large bowl. Add chicken, sliced mushrooms, celery, and onions. Stir together. Add cream of chicken soup, 1 cup water, and remaining spices. Stir well. If mixture is too thick, add enough water to make the filling creamy. Pour into pie pan.

Roll remaining dough flat, and cover vegetable mixture with top crust. Pinch edges together. Using a knife, cut several vent holes in the top of the crust.

Bake for 30 minutes at 400 degrees or until filling begins to bubble.

Dope in a Bottle

No matter the name, there is something special about drinking a dope in a bottle that has a bag of peanuts poured inside.

"Co-Cola," or "Coke," is the name most Southerners use to identify a syrupy sweet, carbonated beverage. Where I'm from, the word "Coke" also identifies Pepsi, Sprite, 7UP, Gingerale, and Nehi Bottling products.

Non-Southerners refer to the same beverages as "soda" or "pop."

For the record, soda is an ingredient Southerners use to bake light and fluffy biscuits. Some folk call the baking ingredient "sodey," as mentioned in this example: "Son, run to the store and pick up a box of sodey so I can make some cathead biscuits for dinner."

The word "pop" evokes a visual image of being popped in the mouth. Perhaps this is how Coke became the universal request for all carbonated beverages here in the South.

The true test of heritage appears when requesting a carbonated beverage. Ask for a soda while visiting or residing in the South, and you are likely to receive a look of pity followed by a slight shake of the head. You might even hear, "Bless your heart, you aren't from around here." Before Coke was available in small towns across rural areas of North Carolina, RC Cola was king. Teenagers flocked to the soda fountain and ordered a dope. Later, when the word "dope" meant something illegal, customers clarified their request by asking for a dope in a bottle.

"Hoke Henderson used to sell dopes for a nickel apiece up there in the Birmingham community," Billy says. "And they shore were good. He'd sell you one for a nickel, or six for a quarter if you agreed to return the empties."

A dope in a bottle is best when paired with a bag of Lance peanuts. Here is the recipe.

(continued next page)

Ingredients

1 ice-cold Coca-Cola (and obviously in the bottle)
1 small bag of salted Lance peanuts

Using a bottle opener, remove the crinkly metal top. Feel sparkles of carbonation kiss your hand. Before adding the peanuts, take one sip from the bottle. If you don't, carbonation will force the peanuts out of the opening.

Open the bag of peanuts and pour them directly into the bottle. Take a moment to watch bubbles gather around the peanuts as they dance inside the glass. Smile. Inhale this life that you have. Even with our struggles, we are blessed. While bringing the beverage close to your lips, listen to the fizz. Drink it in—the sweet, the salt, the essence of joy in a bottle, a remembrance of time gone by.

Cherry Lemon Mountain Dew

In my day, Naber's Drive-in served the best drink in town, a Cherry Lemon Mountain Dew. Hardcore drinkers asked for theirs in a shake cup. Naber's is an authentic drive-in restaurant in Bryson City, North Carolina, that is open six days a week. The place has been around since my parents were teenagers. Roll down the window, place an order through the intercom, and wait for your meal. My personal favorites: a cheeseburger all the way (with slaw), onion rings, and of course a Cherry Lemon Mountain Dew in a great big shake cup. Here is a variation of the delicious drink.

Ingredients

1 Mountain Dew
1 teaspoon of juice from a jar of maraschino cherries
1 maraschino cherry for garnish
1 teaspoon lemon juice

Place 1 teaspoon lemon juice and 1 teaspoon cherry juice in a chilled glass. Add ice. Pour Mountain Dew into glass and garnish with cherry.

Root Beer Float

It's a good idea to keep a couple glasses in the freezer for this delicious summertime treat. One never knows when the urge for a float will strike.

Ingredients
½ pint vanilla ice cream
1 can (or bottle) chilled root beer
1 can whipped cream
Cherries (if desired)

Place two scoops of ice cream into chilled glass, then slowly pour root beer over it. If desired, top with whipped cream and a cherry. Enjoy immediately.

2

Soil Preparation

Growing delicious vegetables literally begins at the ground level. Before dreaming about juicy tomatoes or award-winning watermelons, those who till the soil must first test it to determine where to plant certain seedlings and to assess the amount of supplements the soil requires. Georgia clay, the bane of existence for first-time gardeners, has the potential to produce delicious vegetables when properly prepared and fortified with nutrients. Understanding this, Billy continually feeds his soil organic matter.

Prior to adding organic matter, one should test the area. While soil kits are available at most garden supply stores, many state universities and Cooperative Extension Services offer this service. They also distribute excellent literature and advice on soil enrichment. A computer search will provide the necessary contact information for offices in your local area.

A few decades ago, an extension employee trudged out to your property with a shovel in hand. He did a little dab of digging and a whole lot of visiting. This farmer's friend discussed tips, trends, and livestock prices. Then, without revealing his affiliation, he offered an opinion on the political state of the nation and how legislative decisions trickled down to local farmers. After a cup of coffee and a slice of dried apple cake, two calloused hands met and shook with mutual admiration. The agent climbed into a government-owned vehicle and left with a sack of dirt. These clods would either reveal a variety of deficiencies or provide scientific proof that you possessed the most fertile field this side of the Mississippi.

Usually the results weighed heavy in the deficiency category.

Later, the agent returned with a mathematical chart of nutrients that, when added to the soil, might miraculously convert clods of clay into compost-grade, fabulously fertile soil. The follow-up visit meant another slice of cake, chocolate this time, and the opportunity for a farmer to increase his yield. Science, sweat, and friendship came together to feed America's families.

Now, with limited budgets, decreased staff, and a busy workload, most Cooperative Extension offices distribute an empty box and an instruction sheet. No personal visit. No chocolate cake.

In smaller backyard gardens, the type of soil additives and additional preparation required will vary depending on soil condition and the type of vegetables you desire. Likewise, flowers come with a different set of instructions. While many flowers prefer alkaline, loam, or sand, tomatoes require acidic soil. It is necessary to have an idea what you wish to plant prior to plunking down money for additives that could impede plant growth or encourage disease. Most soil samples suggest the addition of potash, lime, or nitrogen. Consulting a Cooperative Extension agent is worth your time.

Gardening experts often speak about the importance of adding organic matter. For the sake of clarity, when I use the term "organic matter," I do not mean chemical fertilizers. Let me begin with the additive that gets folk in the most trouble, manure.

I live near many horse farms. This unlimited supply of manure is a gardener's dream. When adding manure to the soil, pay close attention to the aroma and texture. I will explain after a quick digression.

Horse farms, especially those located near affluent neighborhoods, are under intense pressure to dispose of equine excrement as rapidly as possible, especially in the summer months when temperatures rise. While the barn smell does not offend me, oftentimes, those enamored with the romantic notion of living in horse country do not understand that animals do, well, smell. A few unhappy neighbors complaining to city officials can quickly transform a serene countryside into a battleground pitting lifelong natives against new homeowners who did not realize that their million-dollar dream home with a pristine view of their neighbor's white picket fence came with the downside of an odor that wafts on the wind.

Looking out their window, newbies observe horses gnawing tree bark and immediately assume that the graceful creatures are malnourished. Uninformed about the mannerisms of the equine, they dial the humane society and report an animal so hungry it eats bark; and, by the way, as long as they have the society on the line, who takes complaints about the smell?

Trouble begins with a complaint and can quickly become a full-on attack, pitting neighbor against neighbor with piles of poo in the middle of the war. Manure has created quite a controversy. It has been the subject of several newspaper articles and the cause of citations given by municipalities. The pressure to offload a seemingly endless supply of waste creates a predicament for equine owners. As urban sprawl creeps into nearby communities, lifelong horse owners have no choice but to relocate.

This is how gardeners receive fresh, unseasoned manure. Most horse farms even offer *free* manure. Some are so eager that they will plunk a scoop into the back of your pickup truck without charge. Unfortunately, most inexperienced, newbie gardeners do not understand the difference between seasoned manure and green (also called hot) horse manure. Adding fresh manure actually damages the foliage and roots of tender plants.

Never apply green manure, from *any* animal, to the garden. So how do you know if manure is green? The answer is a bit complicated. Farmers use the terms "green" and "hot" interchangeably to identify animal waste that has not completed the decomposition process. Incorporating green matter into the soil will either burn delicate plants, cause grass to sprout in your garden, or both.

There are two ways to determine the status of manure: the squeeze test and the smell test. Neither task is for those with sensitive stomachs or a ladylike disposition. Manure that is visibly wet with a slight sheen should remain far away from the garden and allowed to decompose. Organic matter suitable for garden application should smell like mushrooms or dirt, never ammonia. Ammonia means fresh urine, which in high quantity is bad for the soil.

Cedar shavings, commonly used in barns throughout the Atlanta Metro area, are also bad. Manure that is fresh from the stall and has come in contact with these shavings alters soil pH. Do not apply it directly to the garden or near tender seedlings.

Located at the back of Billy's property rests the manure pile, a boxed-in area bulging with both goat and horse manure. Billy firmly believes in adding plenty of manure during the spring. He recently moved a majority of his goat herd into the neighbor's pasture across the road. The homeowners had requested that his livestock live on their property to keep the weeds knocked back and the snakes scared away. Billy's kind gesture came with one significant downside. Because the goats aren't confined to a barn at night, he now has no goat manure for his garden. After he mentioned this need, a nearby horse owner offered what he called "seasoned manure."

When Billy determined that his garden needed horse manure, I was not concerned. This is the same man who cleans out both the barn and the henhouses before breaking new ground. He understands proper compost stages. I was not concerned when he mentioned picking up this load—until he returned and I took a whiff. Literally, the stench of ammonia took my breath.

I also noticed that Billy was exhausted when he got back from the horse farm. I asked if someone had helped him. He shook his head. Immediately, I took issue with the fact that this particular individual (whom Billy was helping by taking the manure off his hands) supervised as eighty-year-old Billy repeatedly heaved the pitchfork until the pickup was full and the bumper had lowered to the pavement beneath the weight of a load that was hotter, and more stinky, than Lucifer's toenails.

Billy casually mentioned, "Now I got to spread the load."

This was his way of asking for help. After I offered my assistance, Billy cranked the Chevy and pointed her toward the barn. When I realized that his plan was to spread the manure one wheelbarrow load at a time, I suggested (begged) him to drive the truck directly into the garden. The thought of pushing multiple loads into the field was one extra step I could not endure. I climbed on top of the pile and tossed the treasures to the ground.

But lifting the pitchfork confirmed my suspicions. Someone had taken advantage of Billy: this so-called "seasoned manure" contained more cedar shavings and equine urine than beneficial organic matter.

That unidentified person had offloaded unwanted content to the first individual available.

Sweet Farmer Billy never gives me the names of people who take advantage of him. He is a wise, *wise* man.

I held my temper and got to work.

"There's no manure in this," I claimed while hurling the contents across my shoulder.

I had insisted that Billy remain seated in the cab of the truck so he could rest. Doing so meant that I shoveled hard and fast, an act that succeeded in getting me filthy and angry.

"This is certainly *not* seasoned," I said with another toss. "Whoever gave you this just wanted to dump it off on anyone. You know, out of sight, out of smell."

Billy shrugged and pressed the clutch to the floor. The truck rolled forward a few feet. "Well, he alleged it was seasoned," he defended. Then he shoved the gearshift in park and grabbed a shovel.

Sometimes it doesn't pay to take a person's word. I held my tongue and tossed the treasures into the garden. Together we spread the manure, even though a feeling of dread pressed heavy on my heart.

Billy seeks the good in everyone. This quality endears him to many. However, the facts were clear. He had been duped, which left us with two choices: distribute the load now, or create a pile at the end of the garden. Billy believed that the soil would immediately benefit from the added enrichment, and, come spring, everything would work out.

I disagreed. Billy may be my elder, but I knew that come summer, we would be pulling grass or running a weed eater through the garden. This is why I never, under any circumstances, add horse manure to my garden. The ratio of grass seed is too high.

"Please. Please don't get any more loads from this guy," I begged. "We do not want this in the garden."

June arrived, and, sure enough, lush mounds of grass thrived between the cornstalks.

"I broke up the soil several times with the tractor," Billy said. "I thought that would take care of the grass."

As I hacked at the earth, I thought, *I could do a little breaking myself.* It does not matter how often you rototill the soil, grass seeds are tough and hardy.

While Billy prefers goat manure, my personal experience has revealed that the decomposition process of goat manure doesn't reach the temperature necessary to kill grass and weed seeds. For me, it is chicken manure or bust.

Each time I incorporate organic matter from the Albertson goat barn, renegade weeds sprout in my veggie garden with the sole purpose of consuming nutrients slated for my tomatoes. These invaders reproduce rapidly and generally tick me off. Some visitors to the farm swear by alpaca or rabbit droppings. Well-seasoned cow manure is also acceptable. Chicken litter, though, is higher in nitrogen. It decomposes at a hotter temperature that kills most weeds. Don't be afraid to experiment until you find what works for you. Never use waste from domesticated animals such as cats and dogs, as both can contain harmful parasites.

Composting is the best way to get adequate organic matter. Many choose to designate an area away from their garden for this purpose. Here they add a variety of items such as grass clippings, leaf litter, manure, and kitchen scraps: like banana peelings or lettuce (never bones or uneaten food, unless attracting raccoons is your intent). After adding items to the pile, water lightly, then expedite decomposition by covering everything with a tarp. Covering the area also keeps unpleasant odors to a minimum.

In order to kill fly larvae and grass seeds, your compost must cook in the sunlight at a temperature that ranges between 120 and 160 degrees. Expect this process to take one to three months in the summer, three to six months during the winter. Decomposition rates depend on a number of variables: the size of the pile, the amount of straw, bedding, or shavings mixed in, and moisture levels. Also contributing is the number of times you have manually turned the pile with a pitchfork. Aeration isn't necessary, but it does increase decomposition. Do not be alarmed when steam rises from the compost pile. Be happy. This is normal. Months after processing, the smelly manure and other items you've composted magically transform into a dirt-like substance called black gold that looks, feels, and smells like freshly tilled soil.

Recycling tea leaves and coffee grounds makes plants happy. Most

Starbucks locations will save used coffee grounds upon request. If you do not have a compost pile, apply both used tea leaves and coffee grounds directly to the soil. Top dressing is acceptable. Tomato lovers should also save eggshells throughout the year. Bake empty shells for two to three minutes in the oven. Crumble into pieces, and then sprinkle directly into the soil. Shells are an easy way to add beneficial nutrients.

The garden is perhaps the most logical place to recycle unwanted newspapers. Several women from the Roswell Garden Club praise the benefits of using newspaper. They claimed that incorporating shredded paper into the soil had transformed their callous, compacted clay into a manageable, workable medium that was capable of growing prize-winning flowers. Since my soil resembles red concrete, I was willing to try anything. I purchased a shredder and got to work.

While many people scatter newspaper into the garden and then apply water to expedite decomposition, soaking the paper before placing it in the garden prevents pieces from flying away on a breeze. After digging a furrow, or as Billy says a "fur" (which is a narrow row in the earth), place the soaked newspaper in the row and cover with dirt. Water well. Once a month, repeat this process and incorporate shredded newspaper into the soil. Fall is the best time to add newspaper. The soil is undisturbed and moisture levels are higher. During the summer, add shredded newspaper in the walking rows and then cover with mulch. Water lightly. Mother Nature does the rest.

What happens next is magical. Worms by the hundreds visit the mushy mess. In a matter of weeks, the paper vanishes. I'm not certain how worms find their way into the garden; I only know that one day there were none, and the next time I checked, hundreds of worms had appeared. Nibbling pieces of the newspaper and munching away on the softened strips of pulp, they grew fat and happy. The worms also left behind their casings (worm poo). Wiggling and crawling through the paper, they aerated the soil and softened the stubborn Georgia clay.

Dirt Cake

It seems fitting that the recipes listed at the end of this chapter maintain the dirt theme. I hope you enjoy these tasty treats.

The dirt cake, a beautiful dessert that will make guests believe you worked hours in the kitchen, is a favorite of garden clubs and social functions. Assemble inside an unused pot and adorn with real or artificial flowers to create a lovely centerpiece.

Ingredients

1 (20 oz) package of cream-filled chocolate cookies
3 cups milk
2 (small) boxes of instant chocolate pudding mix
1 (8 oz) package (light) cream cheese
1 container frozen whipped topping (thawed)

Place cream cheese in a mixing bowl, and mix on low for approximately one minute to soften. Add to the mixture two boxes of chocolate pudding and three cups of milk. Mix well and set aside.

Crumble the entire package of cookies into small pieces. Reserve approximately ½ cup for the top. Sprinkle a layer of cookies in the bottom of the pot, and press firmly. Add pudding mixture, then a layer of whipped topping. Repeat until container is filled. Top with remaining cookies to form a layer of "dirt." Add flowers for decoration.

Mississippi Mud Cake

I don't know if this recipe originates from Mississippi. I only remember this rich cake being one of my favorite desserts as a child. Be advised that the cake portion does not rise like traditional sheet cakes. Also, the dessert is best served warm.

Cake Ingredients
1 cup butter
2 cups granulated sugar
4 eggs
2 teaspoons vanilla extract
1 ½ cups all-purpose flour
⅓ cup unsweetened cocoa powder
2 teaspoons baking powder
½ teaspoon salt
1 cup chopped walnuts or pecans

Frosting Ingredients
1 bag mini-marshmallows
½ cup butter, softened
⅓ cup unsweetened cocoa powder
3 ½ cups confectioners' sugar
½ cup evaporated milk
⅛ teaspoon salt
1 cup chopped walnuts or pecans
1 teaspoon vanilla

Preheat oven to 350 degrees. Spray 9-x-13-inch pan with nonstick spray. In a medium-sized mixing bowl, combine all of the dry ingredients. Set aside.

In a large mixing bowl, combine the butter and sugar. Mix until fluffy. Add eggs and mix well. Stir in vanilla. Slowly add the dry ingredients and beat until well blended. Fold in nuts. Spoon into greased, 9-x-13-inch baking pan and bake for 30 minutes, or until a toothpick inserted into the cake comes out clean.

Turn off oven. Remove cake and add marshmallows to the top. Return cake to warm oven, and allow the marshmallows to melt.

Prepare frosting while cake is cooling by mixing butter, cocoa, sugar, evaporated milk, salt, and vanilla. Combine well and spread over cake. Garnish with nuts.

Butterscotch Haystacks

I remember the first time I tasted a butterscotch chip. Their rich, buttery flavor took our small town by storm and provided an exciting change from chocolate chips. Folk in my hometown called them "Scutter-botches." If it has been a while since you have enjoyed a bag of butterscotch chips, try this recipe. These deliciously soft, yet crunchy haystacks make the perfect snack for Halloween.

Ingredients
1 (11 oz) bag of butterscotch chips
¾ cup peanut butter
2 5-ounce cans Chow Mein noodles
3 cups mini marshmallows

Microwave butterscotch chips and peanut butter together in a large glass bowl on medium high for one minute or until chips have melted. Add Chow Mein noodles and marshmallows. Stir until well coated. Using a spoon, drop onto wax paper. Refrigerate until ready to serve.

Dirty Rice

This high protein dish qualifies as a meal, satisfying hungry farmers like Billy Albertson.

Ingredients

2 cups white or brown rice prepared according to package instructions

1 can beef broth

1 pound Andouille sausage

2 cans light kidney beans, drained

2 tablespoons butter

While rice is cooking, melt butter in a cast-iron skillet and add sliced Andouille sausage. Cook for approximately 5 minutes. Add kidney beans and beef broth. Mix well, then add cooked rice.

Dirty Martinis

Ingredients
6 ounces vodka
Dash dry vermouth
1 ounce brine from olives
4 olives

Chill the martini glass.

Mix vodka, vermouth, and olive brine in a cocktail shaker with ice. Shake until condensation forms on outside of container. Strain into glass and add olives.

Veggie Love

3

Growing "Tow-maders" the Billy Albertson Way

Perhaps it is the fellowship Billy and I share, or our mutual love of the tomato, that converts a simple sandwich into a culinary work of art. Quite possibly, it is the instant gratification, the farm-to-table experience we have when plucking cherry tomatoes from the field. Wiping them on our shirts to remove any dust, we pop the grape-sized morsels into our mouths while continuing our search for larger, sandwich-ready tomatoes.

Suffering through January and February without fresh vegetables makes us appreciate that "first-bite moment." We have waited for the feel of a tomato leaf touching our bare hands as we brush against the greenery. The desire to capture a perfect specimen from which to make an equally perfect sandwich fuels our search. In the grayness of winter, Billy sat in an overstuffed recliner, its arms protected with dishcloths duct-taped to the leather. Warmed by a wood-burning stove, he dozed and rested, dreaming of next year's garden. A few miles away, holed up in my office and chilled to the bone, I wished the heat pump provided the same comfort as oak logs. While understanding that gardens must sleep, I secretly hoped for an early spring so I could get outside and get to work.

The first tomato of the season is highly prized. It's a moment that's been a long time coming. Eager to enjoy the earth's offering, Billy and I have worn a path alongside the rows, nursed delicate plants, snipped suckers from tomatoes, and, while no one was watching, spoken to the vegetables like they are lifelong friends…which, of course, they are.

Making sandwiches is a communal experience. Absent a cutting board, I stand in Billy's kitchen and prepare tomatoes. Juice tickles as it treks down my arm, eventually dripping from my elbow and landing in the sink with a splash. We eat only blemished specimens, reserving the perfect tomatoes for his customers. Billy unwraps a loaf of white bread, lays two slices on paper plates, and then sticks his head inside the refrigerator, searching for the Duke's mayo. With his head still inside, he extends his arm back and places the jar in my hand. Those who have dropped by to purchase a sack of fresh vegetables, and then stayed when invited for dinner, watch our dance, a perfect orchestration of movement inside a tiny 1960s-style kitchen. While I carefully assemble sandwiches on a wobbly center island, Billy emerges with dessert, one of many he keeps tucked behind a container of soymilk.

"Want a slice of cheese with your sammich?" he asks while placing a coconut cake on the counter. Dinner at Billy's always includes dessert.

Shaking my head, I decline. "I'm watching my cholesterol."

A flash of discouragement crosses his face. Knowing that Billy needs protein, I quickly grab a slice of white cheddar and add it to his sandwich. After transferring tomatoes to the bread, a dash of salt and pepper perfects the process. The only thing remaining is a bowed head and heartfelt thanks for the gift of food and friendship.

Tasting the first tomato of the year is a religious experience. The tongue prickles. Awaiting a mildly acidic punch, taste buds wait in anticipation. Inching the sandwich closer to my mouth, my fingers lightly squeeze the fresh bread. Looking across the kitchen table at my lunch mate, I smile, and he smiles, and we both take a bite.

When certified master gardeners gather, conversation focuses on planting procedures, horticulture tips, and proclamations of varieties that produce the most prolific, best-tasting tomato. Each cultivator claims that his or her methodology produces superior results. Conversely, three of the finest tomato growers around—Billy Albertson, Mr. Thomas, and my dad—rarely brag or dole out helpful hints. Instead, they nod, smile, and only answer direct questions.

Learn-by-doing folk are content to keep their hard-earned knowledge to themselves.

Billy's annual planting of the "tow-maders," as he calls them, begins with a trip to D & M Nursery and then to Burger's Market. He has been a loyal customer of both businesses for more than forty years. He visits for a spell, does a little trading, and returns with two or three hundred

Park's Whoppers plants, their leaves waving in the wind as they approach their new home on Hardscrabble Road.

Excited to begin a new growing season, Billy likes to tell people that "a farmer is the biggest gambler around." Given today's rising fertilizer costs and the fuel required to operate machinery, combined with inconsistent weather patterns and perpetual drought, this statement is accurate. Undeterred by the risks, Billy envisions a field lush with healthy crops. He prays while driving, thanking his Maker for the ability to tend the garden at the age of eighty. He believes that idle hands produce ill health, and he can cite friends who died way too soon. This is why he lives every day as if it were his last, giving generously of his time and exhibiting limitless patience to newbies such as yours truly.

He is also in denial about his age.

"Can you believe how old I am?" he said during one of our many discussions about life.

I smiled, having traveled down this familiar conversational path before.

"Of course I don't *feel* eighty years old," he quickly clarified.

"That's because you don't *act* eighty," I said.

Billy laughed. A hearty, head-thrown-back gesture. Coupled with hard work, this sense of humor keeps him young. He lowered his voice a bit and then added, "I've known some people who are old from the moment they're born."

Nodding, I interjected, "Bless their hearts. I guess we were lucky to be born young."

Personally, I think it is neat that Billy doesn't act his age.

After filling a trough with water and adding enough plant food to transform the liquid to a deep royal blue, Billy placed the plastic trays filled with plants into a makeshift wagon haphazardly constructed from untreated plywood and old lawnmower parts. The trough was a gift from a retired farmer, one of many who have sold their property. Sadly, building subdivisions and "McMansions" brings in more money these days than growing summer squash and tomatoes.

Billy steered the John Deere lawnmower across the lawn toward the garden. Nutrient-enriched, reclaimed rainwater sloshed out of the buckets and settled into the bottom of the metal wagon as he approached the crops. Water is a precious commodity. Billy drinks from a well that he drilled fifty years ago when he and his wife, Marjorie, broke ground on their modest home. "Plant water," as Billy calls it, is reclaimed rainwater collected in 55-gallon barrels.

Watching him, I see that he is content on this little strip of land where rainwater saturates the plants, clothes dry on the line, and chickens cluck in their coop. Most who visit, if they are honest, would trade places with him in a blink.

"I've run the tractor across the dirt a few times to break up the clay," Billy said while parking the John Deere. As gardening apprentice, my job was to remove 200 tender plants from their plastic homes and gently drop them to the earth. Billy followed behind with a shovel. As he dug a hole, I doubled back, added a few cups of blue water, then pinched in enough dirt to stand the plant upright.

"Move 'em out just a touch," Billy called in my direction. "You're getting 'em too close."

Regardless of his clear instructions to "take a big step, then drop the plants," my exaggerated steps shortened as I grew tired. Unless equipped with a ruler, plants I plunked had a way of landing too close together, resulting in a jam-packed row of tow-maders. Ideally, Billy preferred three feet of space between each plant. This allowed the installation of cages, manmade mangles of rusty fence so old that pieces literally broke off during placement. I believed that we should install the cages at planting. Billy disagreed.

"I like there to be a little growth on the plants," he responded when I suggested caging and planting simultaneously. While I wanted to get everything into the ground, including the cages and support twine, I rationalized that he knows best.

Besides, it *is* his garden.

One day, a tall, thin man wearing a pair of flawlessly pressed and starched overalls stepped out of an equally spotless Chevy truck and approached Farmer Billy. The two spoke, shook hands, and moseyed to the back of the truck, where the tailgate opened with a clunk. Peering around the men, I discovered a ridiculously large amount of glorious, just-picked tomatoes.

"How much did you bring me today?" Billy asked as the man retrieved a basket.

"Just a few," he replied.

Weighing in at 40 pounds, this tailgate treat benefited both Billy and his customers. Incredibly, when summer demand peaks, 200 plants cannot ripen fast enough to fill the need. This is where lifelong friends help. Carrying their excess to Billy, Mr. Thomas, and others like him, offload what they do not have time to freeze or preserve.

To Billy, meeting the needs of his customers and providing them with fresh tomatoes is important, more than turning a profit.

"I like to keep my customers happy," Billy said. "When my crop is running low, Thomas brings me his excess. That way I keep folk satisfied."

Not to discredit the quality of Billy's tomatoes by any means, but Mr. Thomas grew huge tomatoes that made softballs look small. After the introduction, I vowed to learn as much as possible from both farmers. A full year passed before I secured an invitation to Mr. Thomas's garden.

As he ushered me inside a gated area behind his home, I recognized the invitation as an intimate gift of friendship. These days, not many folk take the time to share their tried-and-true gardening tips. Few allow a personalized tour. Our little stroll was a gift, and I felt thankful.

Mr. Thomas's garden is small, approximately one-quarter the size of Billy's, yet the yield from this area is enormous. When I asked about his process, Mr. Thomas confided, "I use equal parts Super Rainbow Fertilizer (10-10-10), Epsom salt, and pelletized lime."

Leading me down a well-mulched row, he explained, "Each fall, after the garden is finished, my boys dump grass clippings in here. The grass has all winter to decompose. Come spring, we till it into the soil, then we're ready to go."

Comparatively, Billy mulches tomatoes with wood chips and leaves. Regardless, the lesson is clear: successful farmers mulch. They use leaf litter, grass clippings, wood chips—whatever is available. The choice is yours.

"I plant my tomatoes deep," Mr. Thomas said while gesturing with a finger to his forearm. "I'm talking post-hole deep. Then I add a scoop of Super Rainbow, Epsom salt, and pelletized lime. I mix up all three and put it in the hole, add a little dirt, and mix well. Then I bury the tomato in the ground."

Stretching out his thumb and middle finger, he said, "I leave about three inches of the plant sticking out. The rest is under ground. Then I cut off all the leaves except two."

This is where I personally become noncompliant. I am afraid to strip plants down to a nub. Even under the advice of veterans, I do not have the heart to pluck foliage. I knew that other gardeners removed most of the leaves during planting, but until that day, I had never actually met someone who used the process. Mr. Thomas swore by this deep-planting, leaf-removing system. He believed that large tomatoes happened only with a healthy root system.

"The deeper the plant, the healthier the roots," he said while we strolled through the garden.

Mr. Thomas pushed through the plants while I quickly retrieved a small notebook from my back pocket and wrote, *Rainbow fertilizer. Plant deep. Remove leaves.* Tucking the book back in place, I retrieved a camera from the other pocket. Flanked by enormous grapefruit-sized tomatoes, Mr. Thomas smiled. His eyes sparkled. Consciously aware of the gift he bestowed, a personal consultation from someone who had obviously perfected tomato growing, I returned the smile and snapped another photo.

Sensing that I was not convinced, he added, "Snipping the leaves forces energy into the root system. When I pull up the plants in the fall, the roots are three to four feet long."

He opened a pocketknife and lopped off a wayward stem. As he tossed the unwanted piece over his shoulder, an image of rigidity and structure formed in my mind. Billy and I aren't exactly what you call "planners." Instead, we beg. Hope. Coax. Pray that the tomatoes will produce. Here, in this regimented garden, Mr. Thomas expected his tomatoes to produce. Mr. Thomas does his part, and the plants should do theirs.

"It's also important to keep things under control," he said. "I can't have tomatoes falling out of the cages. I keep telling Billy he needs to trim his plants."

Mr. Thomas directed me to a tomato that dared lay a limb outside of the approved growing area. As he whacked the wayward plant, he said, "I don't let any of these get taller than the cage."

Bending forward, he parted the leaves at the base of the plant and said, "Just look at that stalk. See how thick it is?" I leaned in, and my camera clicked as he said, "A thick stalk is the sign of a healthy tomato."

I could almost see the plant respond to his praise.

Approximately the diameter of a quarter, the stalk was thick with a woody texture that resembled a tree limb. Near the bottom, pale nubs formed yellow roots that extended from the sides and plunged into the dirt. This support structure prevented the plant from tilting beneath the weight of immature fruit. It also absorbed nutrients from the soil.

"The moment I see anything shooting out over the cage, I cut it off," Mr. Thomas repeated with another flick of the knife.

Smiling, I added, "My daddy always said, 'a good man carries a pocketknife.'"

Mr. Thomas accepted my compliment with a nod.

"Keeping everything trimmed back focuses energy on the roots and the green tomatoes. It also means a longer growing season."

Continuing to take notes, I confided, "But Mr. Thomas, I don't have the heart. It hurts me to trim the tomatoes."

Cocking his head to one side, he returned the knife to his side pocket and flashed a broad, knowing grin. "Give it a try. What have you got to lose?"

❧

One hundred fifty miles away in Bryson City, North Carolina, my dad works his own little patch of property. There, the March temperatures are cooler than in Atlanta, a nuisance he ignores. Longer days fuel an intense desire to get a little mud on his boots.

Dad's garden is never bare. Months earlier, after the first fall frost, he scattered winter rye seeds, a beneficial cover crop that converts his garden into a lush grazing field for a flock of thirty Ameraucana hens. This greenery eventually fades and returns to the soil, replenishing it with rich nutrients.

Cranking the tiller, Dad walks the machine into the field, determined to incorporate the winter greenery into the earth. Most of the chickens scatter upon his approach, except for one particularly intelligent hen that recognizes the sound as an all-you-can-eat opportunity. Positioning herself dangerously near the metal tines, she gobbles grubs, beetle larvae, and other pests. Soon, her crop bulges with insects.

Propped against a hemlock tree, a plastic container waits. Each year, the local bank hosts a contest for the biggest tomato. Each year, my parents compete. Styrofoam cups filled with leggy seedlings that mother has sprouted also wait for Dad to finish turning the soil. This year, Oxheart tomatoes are their new addition. To ensure proper identification, mother carefully labeled each variety. Unfortunately, the sun faded the ink. Identification will wait until the immature fruit ripens. For now, my parents borrow Billy's terminology and refer to these plants as "wonder tomatoes."

Georgia clay pales when compared to the auburn-colored soil of my parents' garden. Having rich dirt that grows anything doesn't happen overnight; it is created through years of hard work. In the 1970s, Dad capitalized on excess sawdust available at the local mill. He dropped the tailgate and heaved shovel upon shovel of well-seasoned, almost rotten sawdust into the back of his Chevy pickup, then repeated the laborious process while unloading it into the garden. This is one of the main reasons he has a bad back—that and several decades of climbing power poles for a living.

Gardeners often ignore aches and pains, knowing that every inconvenient pain serves a purpose.

After emptying several truckloads of sawdust, Dad would take my brother and me to the bank of the Tuckasegee River. While my brother and I skipped rocks and offered zero assistance, Dad loaded the truck. Heavy with sand, the bumper threatened to spark against the pavement as we returned home. Brother and I received a sandbox out of the trip. The garden received organic matter that adds drainage.

After years of incorporating these additives, Dad has one of the richest garden spots in my hometown. It is so impressive that tourists traveling to the Great Smoky Mountains National Park often stop and snap a photo.

With the soil tilled, Dad places the seedlings in a bucket of water, then returns them to the shade. There are no tomato cages in this garden. Before planting, Dad uses his PhD (post-hole digger) to partially bury several locust stakes in the ground. Creating a row that runs the length of the garden, he wraps a wire around the bottom of the stake and connects it to the next one, and the next. He repeats the process at the top. He then ties sea grass twine to the wire, creating a trellis that keeps his tomatoes off the ground. Only after constructing this trellis does he carefully deposit tender tomato plants into the soil. During the heat of the day, he visits the tomato patch. Opening a pocketknife, he trims the plants and winds the stems around the sea grass twine. Tomato plants are pliable and rarely break during hot temperatures.

There isn't a right way or a wrong way to become a successful gardener. The secret is trial, error, an open mind, and an adventurous spirit. Find a technique that works for you.

ఇ

During the summer, visitors to Billy's vegetable stand know that the early customer gets the prize. He and I hit the field at first light and pick vegetables before the temperature becomes so oppressive and the air so thick that neither of us can breathe. Those who sleep in, or think they can stop by the stand on their way home from work, are disappointed. Billy has learned during his fifty years of farming that "the garden waits for no man." By one o'clock, I am sufficiently exhausted and secretly praying that the vegetable stand is empty so I can head home.

Usually it is.

The economic downturn triggered many Atlanta residents to plant a garden more out of necessity than curiosity. Always willing to offer assistance, Billy welcomes novices onto his little strip of land and, when asked, gladly dispenses advice. Instead of launching into a lengthy diatribe, Billy freely answers questions with a consultative stroll behind the house.

"The best way to learn is by doing," he kindly reminds visitors while escorting them to the garden. "C'mon out back and have a look."

He listens without interruption and responds in a soothing voice that erases their panic. Those who spend time with him leave with this simple man's knowledge and an infusion of confidence that their own property could produce a family-feeding yield—if their homeowners' association allows.

Each year, a number of folk describe issues with their vegetables. The most common disorder is tomatoes with cavernous curlicues circling the top near the stem. Billy acknowledges their concern as he carefully examines the green fruit they have brought that, when conditions are right, should ripen to softball-sized, scrumptious globes of perfection.

Adjusting his glasses, he chooses his words carefully. "What you've got here are water lines."

He explains that as the growing season lingers, weather conditions deteriorate. Soil bakes and becomes hard. The sun, whose face we once sought, beats down, scorching our skin. Heat waves rise from the earth. Rain, if any has fallen, is "hit or miss" and "not much help." As the soil dries, most tomatoes develop thin lines or rings around the top near the stem.

"You're not doing anything wrong," Billy encourages while examining the proffered produce. "Most everyone gets these here water lines."

Billy eases concerns by asking visitors to follow him down the tomato row, where he explains that the imperfections occur during the peak of the growing season, when the weather is dry and the plant is "growing so fast it can't keep up with itself."

If a summer storm drops a torrent of rain after several consecutive days of above 90-degree temperatures, thirsty plants chug much-needed moisture. The peeling, warm and soft from the sun, suddenly expands to accommodate the increased moisture. The result: water lines, which, though visually displeasing, do not affect the deliciousness of the fruit.

Other than maintaining a constant drip of water on the plant, gardeners can rarely prevent this disfigurement from happening, but there are easy rules when it comes to watering plants. First, water at dusk, and never wet the leaves. Whoever started the vicious rumor that nighttime watering causes disease is wrong. When you keep water away from the leaves, plants respond positively to late-evening watering. Irrigating during the early morning isn't advised. The plant doesn't have much time to absorb the moisture. Under no circumstances should anyone water anything during the day. That includes grass. Water lands on the foliage. The sun reflects through the droplets and burns tender leaves. Another must-know tip is that you should not walk on or till wet soil. Doing so destroys soil structure.

"I don't water every day," Billy begins, quickly explaining that he is on a well, not city water. "Keeping the ground wet all the time changes the taste of the tomato and makes the roots shallow. Moist soil can also encourage blight. Now we don't need to fertilize plants during the summer. The plant will put on a bunch of new growth, and the sun will just bake the tender leaves to death."

Blight is a fungus-like organism that thrives in moist conditions and spreads rapidly down the rows. Blight is very bad. Avoid it at all cost. When rain splashes against the dirt, residue coats low-hanging leaves and creates the perfect environment for this disease. Prevent blight by adequately mulching plants with wood chips.

Blossom end rot is another condition that often appears on unripe fruit near the peak of the season. If you notice ugly black spots at the bottom center of green tomatoes, remove all of the blemished fruit and throw them in the trash. Do not place diseased vegetables into the compost pile. Prevent blossom end rot by adding calcium to the soil during spring planting and throughout the growing season. Crushed eggshells, oyster shells, or Tums antacid will infuse the plant with calcium. Using milk from the dairy section is another way to incorporate calcium into your garden. Ask the grocery store manager to reduce the price of milk that is near the expiration date. Stores typically throw expired milk in the trash, but tomatoes love it. Pour up to half a gallon of milk around each plant. Do not add milk to foliage. Pour it around the perimeter. You should see an immediate improvement. If after a week dark spots remain on green tomatoes, remove the entire plant from the garden and place it in the trash.

Having deep-rooted plants begins early in the season. Since roots seek moisture, keeping the top of the soil moist encourages roots to grow upward toward the surface instead of deep in the earth. Shallow roots are disastrous as the days get hotter and the summer drier. Delicate roots that hover near the surface cannot tolerate high temperatures. As a result, plants become diseased and could die.

Despite doing everything within your power to coax delicious tomatoes from the soil, if water lines appear, resist the urge to pluck the fruit from the vine and throw it away. The tomatoes are fine. Think of the plant. It has been in the dirt for many months, enduring temperatures ranging from springtime lows in the 50s to sweltering summer triple-digit temps. By August, the plant is gnarled, exhausted, and practically leafless. Still, it reaches deeper, eager to produce even the smallest golf-ball-sized offering.

Show the tomato a little love. It is one of the hardest-working plants in the garden.

During the summer months, vegetables experience what Billy calls a "frenzy of growth." Even though he sells produce, he also gives away a large portion. When the frenzy arrives at your garden, kindly donate excess to the local food pantry, church, or homeless shelter. Share the abundance with a nursing home in your area. The need is great, and your contribution makes a difference. All too soon, the frenzy subsides and the garden circles back to a period Billy calls "the resting spell."

Removing tomatoes prior to a frost is an easy way to have fresh tomatoes in the fall. Pick unripe produce and place it in a sunny windowsill. Another way to extend the season is to bring tomatoes inside, still attached to the vine. Clip the stems long, and place them in a bucket or small vase of water. The foliage will wither and die, but the fruit will ripen. Change the water every few days.

Before I came into Billy's life, he accepted the first frost as the beginning of winter. He yanked up the vines, rolled tomato twine into an orange ball, and stored everything in what he calls the "plunder building" for the next year. That was then. Today, I keep a watchful eye on the weather and burst onto the property announcing in a panic, "I've got to cover the tomatoes. Frost is coming."

Fall is a battleground between my desire to extend the growing season and the reluctant acknowledgment that Mother Nature deserves a rest. Carrying tarps, bedsheets, and anything else I can find to prolong the season, I drape the plants and transform the garden into a motley display of raggedy bed linens complete with Mickey Mouse sheets saved from my daughter's crib. Using this method, there are usually enough tomatoes for Billy and me to enjoy through December. Even though the harvest is smaller and less juicy than summertime tomatoes, they are more delicious than hothouse, mushy tomatoes.

Take that, Mother Nature.

Eventually, and stubbornly, I will concede to the changing seasons. Until then, I use every technique at my disposal to enjoy fresh tomatoes as long as possible.

The Perfect Tomato Sandwich

The perfect tomato sandwich begins with fresh-from-the-vine, never-refrigerated tomatoes.

Ingredients
1 large ripe tomato
2 slices crispy bacon (per sandwich)
Duke's Mayonnaise
White bread
Salt
Pepper

Prior to slicing the tomatoes, prepare bacon by cooking it in the oven at 350 degrees until golden brown. Drain bacon on paper towels.

In the South, only white bread will do when preparing a tomato sandwich. The nutty taste of whole-grain bread, while enjoyable, competes with the flavor of tomatoes. Use whole-grain bread for chicken salad sandwiches.

Slather Duke's lightly across each slice of bread, place tomato on bread, top with bacon. Sprinkle with salt and pepper. Top with remaining slice of bread.

Enjoy with plenty of napkins to catch the juice dribbling from your chin.

Fried Green Tomatoes

This recipe is for those who have heard of, but never tried, a fried green tomato.

Ingredients
4 green tomatoes, sliced thin
1 egg
½ cup milk
½ cup flour
½ cup cornmeal
½ cup Italian breadcrumbs
2 teaspoons salt
1 teaspoon pepper
Plastic bag
Vegetable oil for frying

Add all dry ingredients to a large plastic bag, and set aside.

In small bowl, mix milk and egg. Dip sliced tomatoes into bowl, then place in plastic bag. Gently shake bag to coat each tomato.

Add enough oil to cover the bottom of a cast-iron skillet. Heat oil. Test oil by partially dipping a tomato into the pan. If it sizzles, it is time to add the tomatoes.

Lower the heat to prevent burning, then add three or four tomatoes. They should not touch, or they will stick together. Fry each tomato 2 to 3 minutes or until lightly brown. Turn over and cook the other side. Remove from heat and drain on paper towels.

Green Tomato Casserole

This is the same concept as a fried green tomato, only baked and with cheese.

Ingredients

4 green tomatoes, sliced thin
½ teaspoon sugar
1 teaspoon salt
½ teaspoon pepper
2 cups breadcrumbs
1 cup shredded sharp cheddar cheese
1 tablespoon butter

Preheat oven to 400 degrees.

Mix dry ingredients in a bowl, and set aside. Arrange one layer of tomatoes in the bottom of a casserole dish. Sprinkle a layer of breadcrumb mixture. Add a layer of cheese then another layer of tomatoes, breadcrumb mixture, and cheese. Dot with butter.

Bake 1 hour or until mixture bubbles.

4

Happy New Year, Southern Style

In the South, those native to the land must consume a traditional New Year's meal. Folk know that when the first day of January rolls around, the menu will consist of black-eyed peas, hog jowls, cornbread, and either collard, mustard, or turnip greens. This tradition is akin to religious doctrine; its importance ranks up there with SEC football and being baptized by immersion. Around here, food equals heritage, and this particular delicacy is a non-negotiable necessity, a tradition handed down for generations and as valuable as Granny's hand-stitched doily.

According to Billy Albertson, his momma always cooked up a mess of black-eyed peas for New Year's. "We didn't have no collards. Momma fixed up a mess of roots and turnip tops. Weren't no hog jowl either; we had fatback and a strip of lean. We ate us some cornbread and turnip roots. Momma boiled the turnip tops in a pot swimming with pot likker."

Mercy, Billy sure does know how to make a girl's mouth water.

"Momma would throw a ham hock into the pot and cook those turnip tops up real good. Then she'd take the grease off the fatback and pour it into the pot. I'll tell you right now, those greens were so good they'd make you want to smack your grandmaw. That pone of cornbread Momma made was pretty good too."

Now before my Northern readers plop a pot on the stove and give these sacred recipes a try, let me warn you that cooking collards, mustard, or turnip tops will release a mighty powerful aroma in your kitchen. Same for hog jowls. One year while visiting a dear friend in the beloved city of Charleston, South Carolina, I took myself for a stroll through the neighborhood. Spanish moss swayed lightly on the ever-present breeze, carrying with it a hint of salt from the ocean and the robust wallop of collards. The fragrance, a medley of bitterness woven with loam, reminded me that no matter where I am when a new year begins, my heart belongs to the South.

Of course, South Carolina is home to hoppin' John, which was, according to lore, so named because South Carolina folk invited guests to the table by uttering the phrase, "hop in, John." The dish goes back to 1841, when it was sold by a black man called "Hoppin' John." A jumble of onions, peas, rice, ham, and seasoning cook to perfection to create this one-pot meal. In the mountains, though, we ate our black-eyed peas like Billy does, with pot likker.

History reveals that in November of 1864, William Tecumseh Sherman, the black-hearted devil, marched from Atlanta on his way to Savannah, Georgia. During that time, his troops stripped Southern farms down to the bare earth. They took cattle, pigs, chickens, and crops that were in the field and stored in barns. Anything worth eating was theirs for the taking. After all, Sherman had a bloodthirsty army of men to feed. Being from the North, the soldiers left fields of lowly black-eyed peas, which up North were called "cow peas" because the plant produced livestock feed. I suspect that leaving Southerners livestock fodder was their final act of cruelty. I imagine the soldiers laughed while thinking, *Let them starve, or graze the fields like animals.*

Except the Yanks underestimated the Southerners who converted the lowly cow pea into a culinary delicacy that even some of my Northern brothers and sisters enjoy today.

Bent but never broken, in 1864, proud folk knelt over the earth and collected the peas that would soon be called "black-eyed peas"—not because the South was sporting a bruising black eye (which it was), but because each nutritious pea contains a single black dot in the center. This harvest would propel generations into a new season of rebuilding and hope.

ঌ

As a child, I hated the traditional New Year's meal, hated the smoke that hovered thick in the kitchen as grease from the jowls crackled and popped in the cast-iron skillet. Hog jowls are, as one might suspect, the cheeks of a pig. It has been said that a pig is a symbol of progress. A pig cannot turn its head to look to the left or right without turning completely around; therefore, it presses forward toward progress instead of looking at the past. We Southerners believe the more fat on the jowl, the fatter your wallet in the coming year. I also despised the black-eyed peas Momma poured out of the can, vowing to eat "just one for good luck" despite my mother's insistence that peas signify prosperity and a single morsel would not suffice. I pressed one pea into a hunk of cornbread, chewed, swallowed, and washed it down with a gulp of sweet tea.

Later as an adult, I visited my friend Linda K's house on January first. My daughter Jamie, who was around eight years old at the time, accompanied me during this drop-in visit. Linda K piled our plates high with peas, cornbread, and a big ole strip of hog jowl she brilliantly called "Linda K's bacon." Jamie hesitated for a moment. She looked at me, then at the plate, then at Linda K.

By the way, if you visit a true Southerner on the first day of January, you *will* partake of the traditional meal. Doing so shows proper breeding and acknowledges our humble rise from the ashes.

"I make my bacon extra crispy," Linda K said while nibbling on a crunchy slice.

Then something akin to a Christmas miracle happened. Jamie took a bite of hog jowl, and another. She licked her fingers and then her lips and asked for more.

&

In the mountains of Western North Carolina, my people don't consume collards on January first. They make an Appalachian delicacy called "Kil't Lettuce." But now that I call Georgia home, I do love collards, my preferred greenery for each New Year. The history of each food is like most Southern lore, based on what your granny told you when you were growing up. Mine told me that black-eyed peas signify coins. At my house, when the phone rang on January first, I knew it was my maternal grandmother asking, "Have you had a mess of black-eyed-peas and hog jowls yet? If not, somebody best git to cookin'."

Greens such as collards, turnip, mustard, or lettuce are the symbol for paper money. Cornbread symbolizes gold. A true Southerner must consume each menu item: black-eyed peas, greens, cornbread, and hog jowl on the first day of January—because my granny and her granny said so.

By the way, in order to achieve prosperity, one must consume 365 black-eyed peas on New Year's Day.

Goodness gracious, Granny, that's a lotta peas!

Kil't Lettuce and Onions

I know the name sounds unappealing, but you're going to have to trust me on this one. I prefer this dish in the spring when the lettuce is tender and the onions strong. Don't worry—this recipe uses olive oil instead of hog jowl drippings.

Ingredients

1 head of store-bought lettuce OR 1 colander full of fresh-picked lettuce (not iceberg)

1 large bunch of green onions, chopped

¼ cup olive oil

Salt

Pepper

Wash lettuce, then chop and drain on paper towel. Place lettuce in glass bowl. Add chopped green onions on top of lettuce. Sprinkle with salt and pepper. Do not toss.

Pour olive oil into cast-iron skillet and heat until oil almost begins to smoke. Once oil is very hot, pour it on top of lettuce to "kill it." Toss and serve immediately with cornbread.

Collards, the Healthy Way

I prefer collards to mustard and turnip greens. Only a trace of olive oil goes into this reduced-fat recipe.

Ingredients
3 slices of bacon baked in the oven until crisp
1 tablespoon olive oil
2 cups chopped collard greens
1 can chicken stock
Salt
Pepper

Add olive oil to cast-iron skillet. Add collard greens and cook on medium. Add chicken stock, then place a lid on the skillet and cook for approximately 15 minutes. Occasionally scrape the bottom to prevent greens from sticking to the skillet. If all of the stock evaporates, you can add small amounts of water. When greens become tender, reduce heat and add pepper and chopped bacon. Taste and add salt if desired. Serve immediately.

Black-Eyed Peas

Soaking dried peas for several hours reduces preparation time and ensures a tender pea. One old-timer has advised me to place a pecan, still in the shell, into the pot while cooking black-eyed peas. He believes this neutralizes the pungent aroma that will certainly fill your kitchen. Personally, I enjoy the smell of a good old-fashioned Southern New Year's meal.

Ingredients
1 cup dried black-eyed peas
1 can chicken stock
2 quarts water (more if beans boil dry)
Salt
Pepper

Sort beans and remove any traces of dirt or tiny pebbles that may be in the bag. Rinse beans and soak for several hours. In a saucepan, add 1 cup of dried beans and 2 quarts of water. Reserve another quart of water should beans boil dry. Cook for half an hour, and add one can of chicken stock. Add a pinch of pepper and salt. Cook for another half hour. Taste. Add more salt and water if necessary. Serve with Linda K's bacon on the side.

Linda K's Bacon (AKA, Hog Jowls)

The secret to perfect hog jowls is baking them slowly. Because the meat will smoke as it gets done, I cook it outdoors on an electric griddle. You can also bake it in the oven, but be advised to use a low temperature and exhaust fan. Microwaving hog jowls is not recommended.

Ingredients
One pack of hog jowls, rinsed
Paper towels

Whether you are using a griddle, cast-iron skillet, or baking dish, it is best to warm the container first. This prevents the meat from sticking to the surface. Hog jowls curl when cooked. For this reason, one should not abandon the skillet during the cooking process. Use a meat press or fork to keep the meat flat and the heat evenly distributed to all areas. Expect to cook meat 3 to 4 minutes on each side, turning often until reaching the desired level of crispiness, just as you do when cooking bacon.

Once cooked, drain on paper towels.

Serve by crumbling into collards or black-eyed peas, or use it like bacon.

5

My Heirloom Tomatoes versus Billy's Whoppers

The second year I helped Billy in his garden, customers requested heirloom tomatoes. In typical Billy Albertson vernacular, he pronounces this particular variety, "hair-loom." He rationalized the need to diversify his crops after the great salmonella scare, which has plagued almost every variety of fruit and vegetable throughout the country. Always willing to satisfy his clients' requests, Billy began dropping the subtle hint, "Looks like I need to grow me a patch of hair-loom tow-maders if I'm going to keep everyone around here satisfied."

Eager to graduate from the title of Billy's helper to real gardener, I volunteered to grow these exotics. Unfortunately, standing between my dream garden and mouth-watering produce was soil that desperately needed inches of organic matter added and more sweat equity than my aging body could manage.

To paraphrase a song from Peter, Paul, and Mary: *If I had a tiller, I'd till in the morning. I'd till in the evening, all over this land.*

Alas, I have never been able to find a tiller designed for my vertically challenged self. Especially not one that I can crank, or one that doesn't buck like a bronco, kick like a mule, and drag me through the garden.

My mother-in-law gave me her tiller, which, after several failed attempts to start the goofy contraption, I promptly took to Billy's. Honestly, Stretch Armstrong himself could not operate a device that requires the user to hold a lever with the left hand, flip a switch with the right, prime a bulb, adjust a choke, pull the start cord, and tinker with a throttle all while simultaneously standing on her head. The last time I checked, humans possess only two hands.

I believe that a legion of tall men manufacture these machines. They are bound and determined to prevent petite female gardeners like myself from achieving horticultural greatness.

It is a conspiracy. I am certain.

After I volunteered to grow hair-looms, Billy studied the soil at my home. He walked the area, bent down, poked, and examined the soil. He glanced heavenward, analyzing the amount of shade, and then said, "I think you need to plant your tow-maders over at my house."

I hesitated at first, knowing that he already uses every inch of ground for his garden. As if reading my mind, he said, "I'll just till up a little patch of dirt," then added, "You can have your own little piece of faller ground at my place."

At the time, I wasn't certain what the words "faller ground" meant, but I could not let this opportunity pass.

While I would love a garden like Billy's at my home, too many barriers exist: shade, compact soil, and the equipment necessary to improve the soil. I need a tractor, a tiller, and a chainsaw, just for starters. Currently, a small raised bed area at my place produces lettuce, cucumbers, and a few peppers. Realizing the futility of growing deep-rooted tomatoes on my shady acre, I humbly accepted Billy's offer.

The offer came without limits. I could grow what I wanted. Let me quickly add that he did not work my area. If I wanted to let the weeds grower higher than my head, or go on vacation and leave the fruit to rot on the vine, that was my prerogative. Billy's soil was my garden to grow whatever I wanted, and what I wanted was tomatoes that tasted like his.

With the success of my first book, we were both under tremendous pressure to reproduce vegetables that were exactly as I had described. Billy was nervous. He realized the need to keep regular customers happy while anticipating a new flock of customers heading his way.

"Zippy, I'm going to expand the patch," he told me, using, as he always does, my nickname that refers to how I zip around and live my life. "Folk are reading that book and coming in here all the time. We've gotta grow more produce."

I quickly learned that people did not know how to respect Billy's property. Folk arrived at all hours. Some missed the message of the book. They were demanding, rude, and pushy. Some arrived on Sunday expecting produce.

"Now I don't do no trading on Sunday," he kindly said. He doesn't pick produce either. God is resting on Sunday, and so does Farmer Billy.

Preparing for the onslaught of readers who wanted to become customers, his daughters and I constructed a variety of signs identifying plants, shrubs, and trees. We also marked areas that were off-limits. Some visitors had mistakenly believed that the farm was a place where children could roam free. All are welcome at Billy's when they observe basic rules: be polite; respect the animals, vegetables, and trees, *and* let Billy rest during the heat of the day (from 2 until 4).

Billy, on the other hand, saw the book as an educational opportunity. "This is my chance to teach city folk how to garden. Kids need to know where their food really comes from," he said.

ঌ

After hooking the sodbuster to the Farmall tractor, Billy steered it toward the backyard, where he made short work of the grass. Metal blades ripped the foliage to shreds and churned chunks of clay to the surface. He made a single pass with the tractor, turned, and then passed through the soil again. Stepping aside as the machinery approached, I was surprised when he turned and made another pass instead of heading toward the barn. Each pass consumed more grass and enlarged the area. After the third turn, my concern grew. Soon my little strip of garden was huge and probably larger than I could maintain.

One of the many lessons I have learned from my time with Billy is that no two people think alike. I have also learned that when it comes to Billy Albertson's offers, one might want to specify the exact dimensions of a little patch. While I envisioned a small compact area in which to cultivate heirlooms, the patch he offered was, in reality, a majority of his backyard.

Have mercy.

After he finished, Billy steered the tractor beneath a clothesline that runs alongside the back of his property. Realizing that he could not clear the cable, I rushed to the wire, stood on tiptoes, and lifted it high enough for him to pass, all while yelling, "Duck!" as he drove the machine underneath.

Billy didn't have a care in the world. He either knew I would hold the metal clothesline safely out of the way, or he maneuvers under it often enough to know he would clear the line.

Grabbing a handy-dandy bucket, I collected clumps of grass that were once the backyard and beat them against a bucket to remove excess dirt. While most urban gardeners would place unwanted lawn litter at the road for curbside pickup, around here chickens serve as a disposal. They play a crucial part of the ecosystem. The moment I began plunking grass in the bucket, excited clucks erupted from the henhouse. Pacing the pen, they pushed and clamored for the best position to access the greenery that held hidden treasures of worms and grubs.

Now I was ready. I thought about how Billy Albertson has farmed the same land for over fifty years. During that time, he has incorporated manure from a variety of animals. Additionally, he has added wood-chips, leaf litter, and grass clippings into the soil. All of this organic matter has decomposed to create soil that will grow practically anything.

And then there was Zippy's just-tilled backyard garden.

The unmistakable sound of a wheelbarrow hitting a rut signaled the beginning of Phase II in the great hair-loom growing competition. Billy had haphazardly stacked three bags of organic fertilizer into a rusty wheelbarrow, which he parked at the edge of my new garden space.

"I thought we'd give this here oar-ganic fertilizer a try," he said.

Having previously expressed my intent to grow organic vegetables, I was secretly thrilled that this load of pre-bagged fertilizer meant we would not clean out the henhouse today.

"I got no eye-dear how much of this here fertilizer to use," Billy admitted as he positioned the wheelbarrow beside me, clicked open a knife, and sliced an opening in the bag. "So let's just be on the safe side and use a lot of it."

Now we're talking.

Heavy feeding is my specialty. I hopelessly apply too much fertilizer to just about everything. Reasoning that plants are always hungry, I apply a little extra boost on a weekly basis. This explains last year's okra disaster.

Six-foot-tall plants.

Zero blooms. The result of too much fertilizer.

Zero okra for dinner.

Billy poured oblong powdery pellets into the bucket. A plume of noxious dust billowed and swirled. Tiny particles attached to my arms. To prevent the inhalation of these particles, I held my breath. After inspecting the bag, I learned that this organic fertilizer was actually dehydrated, compressed, desiccated chicken litter.

Believing that the organic plant food provided adequate nutritional value, we ladled two cups per plant—one in the hole, the other sprinkled as a top dressing after I had tamped the tomato plant firmly in place. I held my breath until I thought my lungs would explode. Turning to gulp fresh air, I thought, *These tomatoes better be delicious!*

While working alone in the garden is rewarding, sweating alongside another soul as chickens scratch nearby is an experience I highly recommend. The sound of a shovel hitting an occasional rock and a rusty can grinding across the bottom of an equally rusty wheelbarrow is the only communication necessary.

Words aren't necessary when the moment is perfect.

ꝏ

So began the tomato competition between the veteran farmer, Billy Albertson, and the wannabe gardener. Billy had the advantage, primarily because I started my seeds late and he had planted four-inch-tall plants months earlier.

Billy confessed, "I don't have the patience to grow tomatoes from seed." He was content to buy plants from D & M Nursery and Burger's Market. Never one for trends, Billy has been supporting small businesses since before it was popular.

The choice between heirlooms and hothouse tomatoes is one of personal taste. Billy's loyal customers preferred his tomatoes, which were either Parks Whoppers or Celebrity depending on availability. I remain puzzled how my tomatoes, which are imperfect, misshapen things of beauty, paled beside his completely round hybrid tomatoes.

I preferred planting seeds without genetic modification, which was why I used Botanical Interests® seeds. Botanical Interests is a family-owned business whose products set the standard for others to follow. Since 1995, Curtis and Judy have supplied gardeners with high-quality seeds delivered in beautiful and informative packets. After contacting a representative and explaining my plan to put their heirloom seeds against Billy's hothouse plants, Botanical Interests® shipped a box containing Beefsteak, Brandywine, Black Krim, Cherokee Purple, and my personal favorite, Ace tomatoes.

Let the competition begin.

The trick to growing humongous heirlooms begins earlier than with hybrids. Depending on the variety grown, some take sixty-five to ninety-five days to bear fruit. Growers of heirlooms must germinate seeds indoors and then transplant early in the season. Planting in June, like yours truly did the first year of the competition, results in a short-lived, low-yield harvest. By comparison, Billy's Parks Whoppers mature early and are dead as a doornail by July.

Traditionally, July and August are the driest and hottest months. One can always irrigate, though my experience proves that plants prefer rainwater to city water.

For best results, introduce heirlooms to the soil immediately after the danger of frost has passed. Cover delicate leaves by placing a plastic milk jug over them at night. Cut away the bottom of the container, and then carefully place the empty jug over the plant.

Gardening isn't glamorous. During the transition from the warmth of your home or greenhouse to the cold, cruel world, these unsightly jugs ensure success. Containers protect delicate plants and help retain heat at night, and they act as a greenhouse during the early morning hours while the sun chases away chilly temperatures. Have plenty of containers available, just in case Old Man Winter and Mother Nature battle. Old Man Winter with his frosty breath and freezing temperatures always wins.

To start plants indoors, use an inexpensive clear plastic container that is approximately the size of a shoebox. Line the bottom with pieces of shredded newspaper. Add enough water to moisten the paper well, then add approximately two inches of potting soil. Carefully sprinkle seeds in rows, and push them into the soil with your finger. Water lightly, then place the lid on top of the box. Place a heating pad under the box to warm the dirt.

Position the makeshift greenhouse near a sunny window. Seeds should sprout in seven to ten days. If the weather is above freezing, take the seedlings outside to harden off in the sunshine. On windy days, leave the lid on to ensure that the temperature remains as warm as possible. The goal is to produce thick stalks and healthy roots.

During my journey as an author and garden club speaker, I have met many wonderful people who have successfully converted their small business dream into a reality. Bill Lucas of Nelson's Grow Best™ is one example. This kindhearted man has a green thumb and a phenomenal product. After personally visiting Billy's garden, Mr. Lucas generously donated samples of Grow Best Liquid Plant Food for me to share with readers who attended my speaking engagements. I am not paid to endorse any product. I use this product, and it works. At the end of the day, people matter. Customers and the service they receive should matter as well. This is why I proudly plant Botanical Interests seeds and feed them with Grow Best Liquid Plant Food.

After watching seedlings turn pale and collapse beneath weak stalks, I placed a call to Mr. Lucas, who explained that my anorexic plants could not support the weight of new leaves. The solution: lightly mist seedlings with Nelson's Grow Best™.

"You can also soak the seeds before planting to give them an extra boost," he explained.

Well now, I hadn't even thought about that.

Mr. Lucas further advised, "The best part about our product is it will not burn delicate plants."

That was all the encouragement I needed. Armed with a spray bottle and determination, I fed the seedlings and the large plants, then watched them respond with lush growth. Smiling broadly, I was confident that my heirlooms were going to give Billy's Parks Whoppers the beating of their lives.

ꝭ

One would think that since my plants were grown within eyeshot of Billy's, they would grow as rapidly as his. One would be incorrect with that assumption. I soon learned that heirlooms mature at a far slower pace than genetically altered varieties. Appearance is another noticeable difference when growing heirlooms. Many varieties produce misshapen, mangled fruit, which has triggered grocery stores to label them "Ugly Tomatoes" so that consumers become accustomed to their appearance.

Shoppers might ask for organic and heirloom produce, but they are often shocked when they encounter them in the store and at the produce stand.

"Folk don't want to take a risk with your tow-maders," Billy said one afternoon as we surveyed our crops.

"But these are the same people who wanted all-natural heirlooms," I defended. "I'm just trying to give them what they want."

"I know," Billy agreed. "People just don't know what they're asking for. They say they want hair-looms, but after looking at yours beside mine, they go for mine every time."

He paused as if weighing what to say next.

"Your tow-maders are scaring my customers."

Well, now the truth was out. People thought my tomatoes—my children whom I had coddled, coaxed, and cherished since germination—were ugly. The only mature thing for me to do was take my tomatoes and go home.

Sensing my frustration, Billy quickly added, "Folk around here...well...they're city folk. They are used to grocery store tomatoes. You know, the grocery store has ruined people for produce."

I pondered this statement while nursing my injured pride. I wondered, *Do consumers really know what fresh produce looks and tastes like?* Most fruits and vegetables are sprayed with methylcyclopropene to slow down the ripening process. Later, an application of ethylene gas induces ripening. Since consumers judge produce by appearance, companies responded by genetically engineering and modifying the food we place on our plates. Commercial farmers are under intense pressure to generate more yield per acre, and do so while praying that they actually make money. They battle disease, drought, politics, and fickle consumers who aren't always educated about the struggle of delivering farm-to-table food.

Farming is dirty work. Picture for a moment my farm clothes. My shirt, my shorts, my hands…even my socks are so dirty they embarrass me. I once pushed a wheelbarrow loaded with tomatoes to the produce stand and encountered a customer who asked, "Did y'all grow these here?"

Did we grow the tomatoes here? Get outta town.

Even the adults these days do not know where food comes from. Moreover, do customers standing in the checkout line realize that two-thirds of all fruits and vegetables are imported? By imported, I don't mean Georgians are enjoying North Carolina apples and North Carolinians are purchasing Georgia peaches. Free trade across state lines is *so* 1960s. Old-timey farmers cringe with the knowledge that the majority of our food travels from distant lands, many identified as being developing countries.

The commercial food industry argues that their produce is as nutritious as organically grown produce. However, no one has compared the nutritional content of vegetables ripped from the earth a week ago and shipped into the United States by freight to that of locally grown produce. Not to preach, but consumers should take the time necessary to learn where food comes from and attempt to purchase locally (or at least regionally grown) fruits and vegetables. Customers should sympathize with the farmer's plight. Municipalities are becoming increasingly strict and limiting what folk can grow on their own property. Comparatively, churches and civic groups that offer community garden space cannot keep up with the requests for garden plots.

Goodness. Let me get off the soapbox and into the kitchen. I have a competition to win.

Billy harvested and pleased his clientele while I picked my beautiful, imperfect tomatoes that I had nurtured from seed to sandwich. At home, I enjoyed the fruits of my labor. I dried them in the dehydrator and canned them using the pressure cooker. The oppressive summer heat lingered. Leaves on Billy's precious Parks Whopper and Celebrity plants curled and turned brown. By August, not a hint of green remained in his rows, but my globes of beauty pointed their shiny green skins toward the hot Georgia sun and ripened to a dark red.

I don't mean to brag, but my tomatoes were still blooming in October while Billy's were lying on the compost heap.

Take that, tomato snobs! Where are your precious hybrids now?

"You got any tomatoes over in your garden?" Billy asked one day. "Folk are getting desperate for a tomato."

"I dunno," I said, "my tomatoes are awfully small. All the large ones are gone." That was true. Once the days get shorter, tomatoes grow at a much slower pace.

"Beggars can't be choosers," Billy responded.

Of course I let him pick my tomatoes. It was his property, after all.

At the end of the season, the results of our tomato-growing competition were inconclusive. Customers preferred Billy's tomatoes during the months of June and July, but I won the taste test at the end of the season. Will Billy stop growing the variety he's grown for fifty years? Nope. Will I stop growing Organic Heirlooms from Botanical Interest Seeds? Not on your life.

I guess we'll call it a draw.

Tomatoes with Basil

Juice from freshly sliced tomatoes blends with balsamic vinegar to form the perfect light lunch or appetizer.

Ingredients
1 large ripe tomato, sliced
1 sprig fresh basil
1 package fresh (whole) mozzarella cheese
2 teaspoons balsamic vinegar
Pepper
Sea salt

Slice tomato and arrange on a plate. Drain water from mozzarella cheese and slice. Place on top of tomatoes. Add fresh basil leaves on top of cheese. Sprinkle with a dash of salt and pepper. Drizzle with vinegar. Serve immediately.

Bruschetta

During the summer, bruschetta makes a light, yet filling addition to any meal.

Ingredients
2 large ripe tomatoes, chopped
1 teaspoon olive oil
2 teaspoons chopped garlic
Pepper
Sea salt
A few sprigs fresh basil, chopped
Crostini or French bread, sliced

Slice bread into small pieces and place in 350-degree oven until lightly brown.

Chop tomatoes and place in a bowl. Add remaining ingredients and toss lightly. Remove bread from oven. Spoon tomato mixture over toasted bread.

Serve immediately.

Summertime Salsa

Summer is the best time to enjoy salsa. Fresh peppers and tomatoes blend to provide a snack so delicious you will want to drink the juice. Roma tomatoes retain their shape when chopped, which is why I prefer them in this dish.

Ingredients
4 or 5 medium Roma tomatoes, chopped
1 sprig cilantro, chopped
1 medium-sized onion, chopped
Pepper (either bell pepper, jalapeno, or habanero, chopped)
1 clove fresh garlic, chopped
1 lime, squeezed
Salt
Pepper
1 bag of corn chips

In a medium bowl, add chopped tomatoes, onion, and peppers of your choosing. Stir in chopped garlic. Using kitchen scissors, cut cilantro into tiny pieces and add to bowl.

Squeeze the juice from one lime over the salsa. Mix well. Add a pinch of sea salt and freshly cracked pepper according to your desired taste.

Serve with corn chips.

Three-layer Dip

Summer is the perfect time to enjoy this refreshing and simple dish. Serve with crackers and enjoy with friends.

Ingredients
½ cup chopped tomatoes
½ cup basil pesto (found in the pasta aisle)
1 package light cream cheese
½ teaspoon garlic salt
1 box of crackers
Nonstick spray

Line a small bowl with plastic wrap, and spray it with nonstick spray.

Place chopped tomatoes in bottom of bowl. Sprinkle with garlic salt. Using a knife, cut cream cheese into wedges and layer on top of tomatoes. Add basil pesto. Fold pieces of plastic wrap over layers and press to remove air pockets. Store in refrigerator until ready for use.

To serve, invert bowl and remove covering. Serve with crackers.

6

Crazy for Chickens and Bunnies

Redbuds unfurl in a glorious pink display. Daffodils line the lawn with butter-colored blooms. Pollen settles in thick layers on decks, porches, and vehicles. Each signals the arrival of spring. Spring is a noisy affair at the Albertson farm. Cries erupt from the henhouse. Wings flap, feathers fly, and the rooster runs himself ragged trying to give each hen equal attention. Instead of an appreciative cluck and submissive stance, the hens "rare back" and threaten to flog his brains out.

Being around a broody hen is not for the faint of heart. Those unfamiliar with animal husbandry conjure an image of pandemonium. Even this analogy pales when compared to Billy's hens. His flock consists primarily of bantam chickens, which he pronounces "banty." Some old-timers call this type of fowl "game hens" or, if speaking about a rooster, a "gamecock."

Known for their talent in the baby-making department, a banty hen will hunker down on a nest and double dog dare anyone, man or beast, to mess with her eggs. She not only refuses to budge but, should a rival hen leave her nest for a bite of food or a sip of water, a determined banty will crane her neck as far as physically possible and steal every single egg from the rival's nest. After rolling an egg beneath her, she assumes a stance that, when translated, means, *Go ahead, try to reclaim the eggs. I dare you.*

My friend, Doreen, often says, "Don't mess with little people. They'll climb your frame in a heartbeat." This saying applies to banty hens. They are smaller and faster than other egg-layers, so you can look, but do not touch their babies.

I adore baby chicks. I must touch, must snuggle, must speak baby talk to the just-hatched fluff bundles. Since being flogged petrifies me, Billy took a hatchling from the momma and then placed it in my hand.

Bringing the ball of fluff to my face, I snuggled, cooed, and quickly returned the chick to Billy, who tucked it back beneath its momma.

Instead of using an incubator, Billy lets Mother Nature select which hen has the privilege of hatching eggs. Billy keeps a close eye on the laying boxes. When he notices the same hen hovering on a nest for several days, he separates her from the regular flock and places her and the eggs into a separate area he calls the nursery. If he doesn't, the hens will fight and end up breaking the eggs.

"Yeah, I had to get the nursery set up," Billy explained one day while gesturing to a large wooden box that was approximately two feet tall. "I've gone crazy for chickens. Some of the girls are in the motherly frame of mind."

"How many hens are setting?" I asked.

"Oh, I reckon I've got ten hens sitting on about seventy eggs."

Lawd have mercy.

Hens in the motherly way hatch in a communal environment. After corralling them into the enclosed area, Billy closes the door and says, "I'll leave it up to them to figure out the nesting arrangements." This habitat ensures daily, if not hourly, fights as hens jockey for the best nest. The first few hours in the nursery are a cacophony of shrieks, wails, and nonstop cackling. Some hens hog the nest. Some refuse to budge while another hen waits, paces, and finally locates her own nesting box. Others quickly yield the right-of-way for more forceful females.

Banty hens are stubborn and prone to bouts of debauchery. I observed one particularly intelligent hen that was determined to hatch her babies with the regular flock next door. Each time Billy gathered the eggs, she hopped off the nest and blended in with the other chickens. She wasn't sitting. No, sir. She had just laid an egg.

Yes. That's right, the warmest one, that's the one I laid. You can have it for breakfast. I made it just for you.

This brilliant biddy would walk over to the feeder, gobble down a couple kernels of corn, and wait for Billy to leave. She had witnessed the hen fights in the maternity ward and decided her babies would not grow up in a home with such discontent. She spied a hole near the most popular nesting box and developed a plan. The clever clucker observed others who were not in the motherly frame of mind, those eager to deposit an egg in the hay and be done. After they laid eggs, she used her beak to pull one egg at a time into a hole in the wall that was so tiny, I wondered how in the world she squeezed through.

Those who study chicken behavior identify this method as building up a clutch of eggs. Common folk like me just smile at her determination.

Billy was worried, though. "I don't know what's going on with my chickens. My girls aren't giving me any eggs. Something must be slipping into the henhouse at night and stealing all of them."

Someone was stealing the eggs, but it wasn't a predator. Carefully tucking egg upon egg beneath her, this hen feigned innocence and, when approached, quickly led us away from her hiding place much like a killdeer does when you approach its nest.

Contrary to vicious rumors, chickens and other feathered friends are not stupid.

Billy's aggressive broody hens don't care one bit who knows that they are in the baby-making business. I don't need to enter the henhouse in order to feel their wrath. A broody hen is ill, inhospitable. She greets anyone who approaches with a sound that resembles a growl. She hurls threats and backs them up with action.

This particular hen was the best momma because she would fight a bear for her biddies. Gazing upon Billy with disdain, she prickled her feathers. She did not want to move into the maternity ward. No way. She rared back, ready to peck, launch, flog.

"Now now," Billy said calmly. "I'm just moving you to the nursery."

The moment he separated her from the unhatched eggs, she cackled and cried. She paced the cage and poked her head through the openings in the chicken wire. She knew she belonged with her babies. Why didn't that human man understand?

Reaching into the hole, Billy gathered the still-warm eggs in a bucket and put them in a nest. The renegade mommy did not approve. The moment he stepped away, she gently entered the nest, careful not to damage the eggs. Using her beak, she turned the eggs until they were to her liking. She slowly eased herself over the center of the nest and lowered her body until the eggs disappeared beneath her soft feathers. Then she assumed the look that dared anyone to approach. Approximately twenty-two days later, baby chicks emerged and the broody hen bounced off the nest, her beak held high, proud of the tiny babies trailing behind her.

ꝏ

I have always had an intense desire to own bunnies. There is no rational reason behind my bunny obsession other than they are cute, cuddly, and produce chocolate cream-filled eggs for Easter. My husband does not share my passion or adoration. But his plans for our home to remain a bunny-free zone ended when fate and destiny determined otherwise.

Following the unexpected death of my dear friend Mr. Ora Coleman, a fellow animal lover named Andrew Wordes offered temporary housing to all of Mr. Coleman's animals. A few days later, Andrew called to say, "I need to find Mr. Coleman's bunnies a permanent home. They come with forty pounds of food, water bowls, everything you need. Do you know anyone who would like to adopt them?"

I most certainly did. I knew three of the biggest bunny lovers around: me, myself, and I.

While we were dating, I had warned my husband that I am an unpredictable gal who is prone to spontaneous events. I offered examples, saying, "If you go out of town, you might return to a repainted bedroom or a rearranged house."

He had smiled and said, "That's what I love about you, never a dull moment."

Life was about to become interesting.

The bunnies came with a McMansion-sized cage that allowed plenty of room for expansion. Constructed of quality wood and placed on hand-carved legs. Instead of being three feet off the ground, this home positioned the rabbits only one foot from the earth, but they were still safely away from predators. Even though fortified with wire and structurally sound, the cage needed a few necessary, extra reinforcements to keep the bunnies safe from raccoons and coyotes. I had waited ten years for these cottontails and would not allow a masked bandit to snatch them in the middle of the night.

To discourage any varmint intent on eating my babies, I hammered a row of u-shaped nails two inches apart along the bottom and top of the structure. While blood dripped from my hit-with-the-hammer fingers and smeared into the white paint, staining it an effeminate pink, the bunnies waited in a pet carrier, contently munching carrots.

My newly adopted bunnies were angels. I had witnessed their peaceful personalities during my visits with Ora Coleman, Roswell's celebrity farmer. Mr. Coleman loved all of God's creatures, especially Smartie, his mixed-breed canine who sealed our friendship with a cat-like rub against my leg.

"I call him Smartie," Mr. Coleman had explained upon our introduction, "because he is so smart. He don't usually like people, but he sure has taken a liking to you."

Smartie stood sentinel at a door scarred with scratch marks. Mr. Coleman first attached the dog to a lead and then opened the door. Smartie bolted outside and marked his territory.

"Look at that Smartie," Mr. Coleman said while admiring the animal no one else had wanted. "He sure takes good care of me."

Mr. Coleman clasped his hands behind his back while I followed him up the hill. The Coleman Farm is a serene spot nestled in the heart of Roswell, Georgia—a place where Mr. Coleman's momma and poppa once raised everything they needed to survive. Mr. Coleman preferred the company of animals to humans. At his place, chickens scratched, turkeys gobbled, and bunnies enjoyed the good life with little human interaction. They wiggled their precious bunny noses and nibbled proffered carrots while Mr. Coleman spoke softly, then carefully eased them from their cage. I stroked their fur while he said, "Isn't this just like heaven?"

For those who, like me, have not raised rabbits, I submit that they are patriots of peace. Petting a bunny is a calming process that releases stress. You can almost watch waves of tension leave your body. However, let me offer a few words of caution. Bunnies, even tame ones, can climb your frame and scratch the living daylights out of you.

Back home, I opened the pet carrier. Explaining in a soothing voice that "this is your new home," and "your daddy is in heaven now," I reached inside. "Let's check out the new home Uncle Andrew gave you."

At that moment, the largest bunny grunted and launched forward, firmly intent on ripping my hands to shreds. I was in shock. After ten years of begging for bunnies, I found that Mr. Coleman's children hated my guts.

They're traumatized, I reasoned. *They have lost their daddy and are in their third home in three weeks*. Nodding my head, I affirmed that I was a good momma and that the bunnies just needed an adjustment period.

The problem of how to transfer them from the cramped carrier cage into the bunny condo without me losing a limb in the process remained. Eventually, I pressed the carrier to the condo, opened the doors of both, and watched them hop into their new home.

I didn't immediately tell my beloved husband about our newest additions to the family. In fact, I hid their home under our deck behind the house. A week passed before I had the courage to come clean.

"Honey. We need to have a conversation," I began while motioning for him to have a seat.

Turning to face me, he waited for what came next.

"I've done something bad."

No response.

"Something really, *really* bad."

Usually, a confession that begins with the phrase, "I've done something bad," spells trouble. Vows broken. Truth shattered. Irreparable damage.

Not in my marriage.

"We have bunnies," I announced. Speaking rapidly, I added, "But don't worry. They are male bunnies. They came with fifty pounds of food and a house. They are Mr. Coleman's bunnies. It is an honor, really…an honor to be asked to care for Mr. Coleman's bunnies. You know how much he loved them. Did I mention that the smallest bunny only has one eye?"

His response: "Let's go have a look at them."

My beloved wasn't exactly pleased, but he does practice tolerance.

In lieu of an olive branch, I offered a carrot, which the largest bunny took with grunts and lunges. In the days that followed, the bunnies never warmed to me. I had hoped they would at least be thankful. They were not.

"Perhaps they'll appreciate the opportunity to graze," I said to my daughter, Jamie. "Let's turn them out in the yard."

After converting a metal cage into an outdoor grazing pen, I announced during dinner, "I'm putting the bunnies to work. They'll earn their keep and keep the grass trimmed."

My husband was skeptical.

Jamie, though, has a way with animals. She had also begged for bunnies, which was why I happily transferred ownership to her. Each day she placed them in their outdoor grazing pen.

"See," I said. "Pretty soon we won't even need to cut the grass."

This might have been the case, were it not for the fact that bunnies, even domesticated ones, love to dig holes. *Deep* holes.

"I don't have good luck with bunnies," Billy shared when I told him about my trouble. "People are always bringing me cast-off animals, especially after Easter. Someone brought me some bunnies once, and they lived about a week before dying. I was cutting the grass around their cage, and next thing I knew both of them keeled over dead. I reckon the lawnmower was so loud they had a heart attack."

Today, Mr. Coleman's bunnies remain on my property with Jamie as their caretaker. When she leaves for college in a few years, I guess I will hire a nanny.

No Fail Hard-boiled Eggs

This is the easiest way to cook perfect hard-boiled eggs every single time.

Ingredients
6 medium eggs
Water
Saucepan

Place eggs in saucepan and cover with half an inch of water. Bring water to a boil. Once water begins to boil, cover saucepan with a lid. Turn the heat off. Let eggs sit in hot water for 10 minutes. Remove eggs and place them in cold water for 5 minutes. Crack, peel, and enjoy.

Deviled Eggs

Deviled eggs are a staple of any Southern gathering. Quick and inexpensive, these disappear fast.

Ingredients
6 hard-boiled eggs
2 tablespoons mayonnaise
Dill pickles
1 teaspoon salt
½ teaspoon pepper
2 teaspoons prepared mustard
Paprika

Slice hard-boiled eggs lengthwise. Remove yolks and place in a small bowl. Using a fork, mash yolks. Chop one dill pickle into tiny pieces. Add to yolk mixture. Stir in mustard and mayonnaise. Add salt and pepper. Note: If mixture seems too dry, add small amounts of pickle juice to the yolk mixture.

Spoon mixture into white egg halves. Sprinkle with paprika. Serve chilled.

Stick-to-your-ribs Omelet

Ingredients
Nonstick cooking spray
2 large eggs
½ cup shredded cheddar cheese
¼ cup sliced mushrooms
¼ cup chopped onion
¼ cup green pepper
¼ cup sliced ham
Dash hot pepper sauce (optional)

Spray cast-iron skillet with nonstick spray. Let skillet warm over medium heat.

Mix eggs and cheese. Add salt and pepper.

After skillet is hot, pour egg mixture into pan. Allow to cook until it is no longer runny. Then add remaining ingredients.

Using a spatula, carefully fold the omelet in half. Serve immediately.

Egg Salad Sandwich

Ingredients
2 hard-boiled eggs
2 slices bread
2 large dill pickles
1 teaspoon Duke's mayonnaise
1 teaspoon prepared mustard
Salt
Pepper

In a small bowl, chop hard-boiled eggs into small pieces and sprinkle with salt and pepper. Chop pickles into tiny pieces and add to bowl. Stir in mayonnaise and mustard.

Toast bread. Apply mixture to bread.

Slice remaining pickle lengthwise in thin slices and add to sandwich, or use as garnish.

7

Kudzu, the Vine that Ate the South

What I am about to share has previously been marked "classified." Only those with a super-duper high-level security clearance (meaning myself and Jamie, an accomplice to the debacle) have had access to this data. Since we are all family, I will tell y'all what happened. First, you must swear an oath. A pinky promise, spit-in-your-palm-and-shake-on-it, cross-your-heart, needle-in-the-eye affirmation that you will not judge us.

Before I begin, let me attest that no animals were injured or maimed during the debacle. Jamie and I, however, are scarred for life.

It began with kudzu.

For those who don't know how kudzu arrived in the South, let me start the story there. First introduced at the Japanese Pavilion of the Philadelphia Centennial Exposition in 1876, the plant wooed unsuspecting victims with lovely aromatic purple blooms and lush leaves. The vines wrapped tender tendrils around folks' hearts. Enamored by its enticing fragrance, homeowners planted the creeping, ground-covering plant alongside the peaceful porches of their majestic white-columned Southern homes. They were blissfully unaware of the tragedy about to befall them. The vine grew, bloomed, and quickly became an unmanageable shackle whose tender tendrils hardened like a vice, squeezing the life out of most everything it touched.

Ignoring warnings from botanists, the United States Soil Conservation Service advocated the species. In the 1930s, they determined that kudzu could control erosion, improve soil quality, and feed livestock. The government identified kudzu as a miracle plant and paid farmers $8.00 per acre, a fortune in the '30s, to grow the vine on their property. By 1946, 3 million acres of farmland hosted this fast-growing cover crop. Four short years later, the federal government reversed its stance and no longer recommended the planting of Kudzu.

By then, my dear friend, it was too late.

In the 1970s, the miracle plant crept into the weed category. The facts about Kudzu were widely known at that point. With roots reaching 12 feet deep and snake-like tendrils that truly can grow a foot in a single day, the vine is now viewed with disdain by Southerners. In 1997, a noxious weed designation clearly conveyed how the South felt about this non-indigenous gift from our Asian and Philadelphian neighbors.

Kudzu is like the crazy aunt every Southern family has. Unmanageable. Out of control. Something we'd like to ignore but can't.

Today, covering more than 7 million acres of southeastern countryside, kudzu is called "the vine that ate the South." And with good reason.

ቖ

Hidden far away from annoying homeowner associations, my Atlanta home was the closest thing to a mountain cabin that we could find in the heart of downtown. In addition to ridiculously high property taxes, this serene setting came with the price of rope-like tendrils of kudzu spawned by Lucifer himself, intertwined with another horticultural annoyance from Hades, the wild grape.

As with most disastrous events, that particular day began innocently. My daughter and I were at Billy's when I commented, "The kudzu and wild grapes are taking over my place."

Ever helpful, Billy said, "I got just the thing you need."

Since Billy's place does not grow even the tiniest tendril of kudzu, I believed he would suggest a weed killer that was one drop more powerful than Roundup and a tad less traumatic than Ground Clear.

"Oh, yeah," he said with a nod toward the goat pasture, "I've got just the thing. I'll let you borrow old Hornless. He'll knock back your problem in a day, maybe two."

Hornless was a compact, low-to-the-ground creature with a dirty blond mane and a strong, square chin. He'd arrived at the Albertson farm like many other animals; someone could no longer care for him.

"Folk 'round here use my place as a dumping ground for their animals," Billy said with a smile.

It seems that city folk romanticize about having a slice of country, only to face the reality that farm animals, even those categorized as pets, come with tremendous responsibility.

A lesson I soon learned.

Gazing upon Hornless the goat while he munched contently in the pasture, I realized that he had to be the solution to my kudzu problem. He alone could resolve the issue in an economical manner without poisoning the environment or costing me a fortune.

ღ

Hiring goats as grazers isn't a new idea. Municipalities such as Chattanooga, Tennessee, ignoring guffaws and jokes made at their expense, have enlisted certified Goat Browsing Contractors who offer small herds to private property owners. For the low price of $50.00, plus a $10.00 administration fee, homeowners in Chattanooga select from a list of bonded contractors who install temporary fencing and, if necessary, assign a llama to stand guard against hostile dogs and coyotes while the contracted goats nip weeds clean down to the dirt.

Before you giggle, you should know that the Hartsfield-Jackson airport employs both sheep and goats. I've even spied at least a dozen goats grazing the hillside as part of the Army Corps of Engineers' greener grass mowing initiative. I guess you can say that the goats provide a literal application of the term "weed eating."

ღ

"All you need is a chain and a collar," Billy said, then quickly added, "He came with a collar, so you're halfway there."

Not far from my home, three pygmy goats manicure my friend Kelle's lawn. Secured on a lead she has corkscrewed into the ground, her goats constantly groom the grass. When Kelle's three boys play ball in the yard, the pygmies also participate. One could say she has six children. Thinking about Billy's offer, I reasoned that if Kelle manages three goats in the middle of a subdivision, surely I could handle one tucked deep in the woods far from sight.

Seeing the BABY GOATS sign at Billy's remains a begging point for my daughter who, even as a teenager, still pleads for a four-legged companion. She had lobbied, begged, and offered to do *anything* for a goat. She had waited three long years to hear me say "yes."

Which I quickly clarified when I said, "We are only *borrowing* Hornless for a couple of days."

Jamie squealed and gave me a tight squeeze.

The sales clerk at the pet supply store directed me to a well-stocked fencing aisle that featured a variety of restraints for many large dog breeds. I soon located a Spiral Take Out Stake made by TopPaw™ with a 30-foot cable that boasted 920 pounds of break strength. Placing it in my basket, I stood in the checkout line, envisioning a beautiful Martha Stewart-like lawn. The directions on the package were simple. *Just screw into the ground. Attach the cable, and you're ready to give your dog room to roam.*

Surely what's good for the dog must be good for the goat. How much easier could life get? Yes, I imagined with a confident lift of my chin. I could do this. I would take the spiral stake home, twist it into the earth, clip good ole Hornless to the lead, and then sip on some sweet tea while watching the kudzu disappear.

Turning the package over, I read the two magic words all goat owners want to see, *Super Strong.*

Eradicating kudzu at my house would be super easy.

Billy and I pushed, turned, and eventually pounded the stainless-steel tine until it wedged firmly into the ground. He then clipped the lead to the cable, made certain Hornless was secure, and steered his truck home. I went inside to pour a glass of tea, confident that my environmentally clean mower was hard at work. Minutes later, I stepped outside to inspect the area. Already planning the sales pitch I needed to convince my husband that Hornless was a worthy addition to the family, I searched the area where Billy and I had left him.

Imagine my shock when I discovered that attached to the end of the *Super Strong* cable was the super weak plastic clasp.

Hornless was on the loose.

Rushing inside to grab a bag of carrots and fill a plastic cup with dog food, I yelled for Jamie to help me corral the escapee. Jamie shook the cup as Billy does to signal dinnertime.

Hornless was not fooled.

After we had chased the beast around the house for over an hour, Hornless rested near the swing set. Snatching the carrots from my hand, Jamie shook the cup and dangled the bag.

Hornless did not care.

Establishing a wide perimeter around the goofy creature, we flanked him and closed in for the capture. With a confident lunge, I jumped forward. Soon the goat would be back where he belonged, at Billy's house.

Easier said than done.

With a low center of gravity and a determined burst of energy, Hornless bolted past us. Hooves clicked down our driveway while I watched. A helpless feeling of impending doom punched hard in the pit of my stomach.

"We've got to grab him before he reaches the main road," I said.

In order to catch a goat, one must think like a goat. Clearly, Hornless did not want food. Reasoning that I should step away and allow him time to acclimate to his new home, I motioned for Jamie to retreat and give him some space. Instead of calmly exploring the area and returning to the woods, he saw our retreat as an opportunity for freedom. As he scampered further down the driveway, turned right, and clomped double-time fast down the yellow centerlines of the main road, I realized the gravity of his escape.

Rushing to the street, we watched his getaway. Powerless to stop him (he obviously would not come when called), Jamie and I joined hands and hoped for the best. I knew that using my car to give chase was futile. How could I catch him? What could I do? Seconds later, a Good Samaritan pulled over and pressed a cell phone to her ear.

It is remarkable how fast the city's finest appeared after someone reported a goat trotting down the middle of the street. Honestly, Miss Good Samaritan was still on the phone when a flash of blue lights confirmed it was time for Jamie and I to do what was best. I am not proud of what happened next. My daughter and I fled the scene. Weighing the probability of arrest, I determined it was high time to hightail it out of there and hide where no one would ever think to look for me…the shopping mall.

Casting a final glance in the rearview mirror, I witnessed others attempting to guide the lost animal home. I reasoned that, if left alone, Hornless was just a few yards from the City Park, where he could munch happily on kudzu forever and ever, amen.

While some may judge and think my behavior irresponsible, many will sympathize. They have felt the same punched-in-the gut panic after an animal broke through the fence. They have chased goats. They understand, like I did, when it was time to walk away and let the creature settle down.

Actually, when I thought about it, having Hornless maintain the woods would save the city money. Yeah, let's go with that idea.

Opting to leave the past in the past, Jamie and I learned our lesson: goats belong at Billy's, not in our backyard. We vowed never to speak about the incident. Additionally, we determined that we would wait a few days before telling Farmer Billy we had lost his goat.

Allow me to introduce Mr. Hal. Like Farmer Billy, Hal lives in an area that was once farmland. Surrounded with subdivisions, his charming property features an enormous oak tree equipped with a swing, and a small pasture full of goats. The day I interviewed him, he was celebrating the arrival of almost a dozen babies.

"I've got too many goats," he confided while surveying the field. "I need to get rid of a few."

Boy, if I had a nickel for every time I've heard that statement. Apparently, goats reproduce like bunnies.

"You know someone once stole one of my goats," Mr. Hal said, "and they brought it back a week later."

This provided an opportunity for me to ask, "Do you remember someone bringing you a blond goat a while back?"

Mr. Hal thought for a moment then said, "I remember that goat." Pointing to the pasture where a long-haired, multi-colored goat grazed, he asked, "Did he look kinda like him, only without horns?"

I nodded. "Yes. Except he was a bit slimmer and much shorter."

"Wasn't that Billy Albertson's goat?" Mr. Hal asked innocently.

I nodded again.

"What I never understood is how he got inside my pasture."

Puzzled, I asked, "You mean a police officer didn't deliver Hornless to you in the back of a patrol car?"

A smile formed as he reached up to scratch his chin. "No. I just looked out the window and there he was. I always wondered how in the world he got inside the fence."

Hmm. I wondered that myself.

After breathing a sigh of relief, I felt that I was finally able to share details of the escape, our search, and, ultimately, our retreat. Mr. Hal chuckled.

"My daughter and I lived in fear that the police would arrest us," I admitted. "I figured Hornless would find his way back to Billy's somehow. Weeks later, I drove by your house on the way to the library and saw a sign on your fence that read, GOAT FOUND. I knew the odds were slim that anyone else had lost a goat. By then I had confessed to Billy (not my husband) that Hornless was lost and roaming the woods. So I called him and said, 'I think I've found your goat.'"

Displaying a smile that comes with personal experience of wayward animals, Mr. Hal said, "It all worked out for good. We ended up doing a little goat trading."

Mr. Hal pointed across the road and said, "Ya know, I once had this old bull get out. One day I got a call from the man who lives where Target is now. He said, 'Mr. Gronholm, do you have a cow out?' I looked at the pasture, did a quick count, and sure enough, the old bull was missing. So I grabbed the feed bucket and a rope."

To catch a bull, one must think like a bull. Incorporating my goat-capturing methodology, Mr. Hal described a similar attempt to capture an animal that did not wish to return home.

"I figured I'd feed him a little grain, and then use the rope to lasso him."

For the record, stories featuring a bull and a lasso rarely end well.

"So I fashioned the rope into a lasso and laid it around the bucket. Then I shook the bucket enough to get the bull's attention. The bull stuck his head in the bucket, and I pulled the rope tight. After that, I figured all I needed to do was walk the bull back to the pasture. Shouldn't be no trouble at all."

Mr. Hal smacked his hands together, clapping so loudly that I jumped. "That old bull took off like a shot. He lit out so fast that the rope hissed and smoke starting boiling off it."

Envisioning the beast dragging Mr. Hal down the pavement, I asked, "You didn't have the rope tied to your arm, did you?"

Even though he shook his head, there was something about the way he looked down at his hands and lightly touched his wrist that made me think otherwise.

"Did someone call the police on you like they did me, or did your neighbors help?"

He chuckled. "That was before Highway 92 was six lanes. Back then, when animals got loose, the police would come stop traffic, even help if need be. Naw, I had to call someone with a horse trailer who helped me load the bull."

As Mr. Hal and I walked the perimeter of his property, a peaceful feeling filled my soul. This self-proclaimed retired gentleman owns a lovely piece of property that, over the years, has been whittled down to three manageable acres inside the city limits. His home is located on one of the major thoroughfares in the area. A small herd of goats call his place home. He welcomes visitors, encourages folk to line up and toss food to the goats. He receives so many visitors that the goats have their own Facebook page, "Gronholm Goat Farm."

"It is great to live in a city with leaders who welcome this type of mixed use," Mr. Hal said. "To some, goats may be a nuisance; to others they are a teaching opportunity. Just the other day a Mercedes pulled up and a lady stepped out of the car. She kept a tight grip on her daughter's hand the whole time."

Mr. Hal's chin quivered. He stopped speaking for a moment and cleared his throat. Then, whispering to himself more than speaking aloud, he said, "I'll try to get through this story, because it is important. People are always stopping by. On that day, the mother asked me if her daughter could see the baby goats. I told her 'of course.' When I looked at the little girl, you could tell she was special needs…that a strong grip was necessary to keep her from dashing into the road and getting hit by a car."

Mr. Hal looked away and wiped tears from his face. I averted my eyes to allow him a private moment. "That's why I keep the goats," he said, "for the children. Young children need access to animals.

"Last week when the baby goats were born, there was a huge crowd. All the babies came at once. Folk lined up with their cameras. It was quite a show." Mr. Hal's eyes sparkled with mischief. "No one has to explain where babies come from to those children. Children of all ages and abilities need a place to learn about animals. My place does that for many people."

Mr. Hal had offered me a glimpse into his soul. It was my turn for tears.

As a light rain fell, we continued our conversation inside. I soon learned that good ole Hornless wasn't the only animal to mysteriously appear in Mr. Hal's pasture.

"One day I looked out the window, and wouldn't you know there were two horses standing in my front yard. I put them in the pasture, then called everyone I could think of to see if they were missing a horse. No one was. So I fed them for three weeks. Then one day, out of the blue this guy pulls in with a horse trailer, steps outta the truck, and says, 'You got my horses.'"

The window from which Mr. Hal gazes seems filled with adventures.

"And he was not a nice man, talked to me like I had stole his horses, didn't say where he was from, nothing. Heck, even his horses didn't like him. He had a time loading them. His horses acted like they didn't want to leave here. They certainly didn't want to go with him. But when he did finally get them in the trailer, he just slammed the door and drove off without saying a word. Not pea turkey squat."

Interrupting, I said, "Wait a minute. This man didn't say thanks, didn't offer to pay for the feed, or any damages the horses might have done to your pasture?"

"Nope," Mr. Hal said with a shake of his head. "Didn't say pea turkey squat. Just slammed the door and drove off."

Sitting beside the fire, we exchanged more stories of runaway animals, bird watching, and the joy of fly-fishing. During our talk, I noticed a sign above the piano that read,

MOTTO TO LIVE BY:

Life should not be a journey to the grave with the intention of arriving safely in an attractive and well preserved body, but rather to skid in sideways chocolate in one hand, martini in the other, body thoroughly used up, totally used up, totally worn out and screaming Who Hoo what a ride!

Can I get an amen?

Controlling kudzu is no longer a priority. I still pull it up with my hands. With Hornless the goat returned to Billy and a newfound friendship with Mr. Hal, I anticipate many adventures ahead. I never imagined that my adventures with Hornless years ago would bring a new friend like Mr. Hal into my life. Perhaps soon he will teach me how to fly-fish. He promised he would, and I believe he is a man of his word. I warned him that he must unhook all the fish I catch. Not because I'm a squeamish girl, but because that is my strategy to reel in more than this expert outdoorsman.

Fly-fishing…now there's something I could do. How difficult could it be?

GORP

Hikers the world over recognize this acronym as Good Old Raisins and Peanuts. At my house, it stands for Goats Out Roaming the Property.

Ingredients
½ cup semisweet chocolate chips
½ cup dried figs, chopped
½ cup dried blueberries
¼ cup sunflower seeds
½ cup salted almonds or peanuts
½ cup granola cereal

Mix all ingredients and store in airtight container. Enjoy as a snack, or sprinkle over yogurt or hot oatmeal.

Disappearing Zucchini and Squash

This dish is so yummy, it will vanish almost as fast as Hornless.

Ingredients
1 large zucchini, sliced
2 small squash, sliced
½ cup uncooked orzo pasta
1 can chicken broth
1 large onion, chopped
1 garlic clove, chopped
Dash of pepper
Pat of butter
¼ cup shredded Parmesan cheese

Prepare orzo according to package instructions, using chicken broth instead of water during the cooking process.

While orzo is cooking, slice squash and zucchini. Chop onion and garlic. Place a pat of butter in a cast-iron skillet. Heat until butter has melted. Add zucchini, squash, onion, and garlic. Sprinkle a dash of pepper. It will not be necessary to add salt. Chicken broth and Parmesan cheese usually contain enough sodium.

Sauté vegetables for 3 to 5 minutes until they are the texture you desire. Drain orzo and place in bowl. Stir in vegetables. Sprinkle with parmesan cheese. Serve immediately.

Escape Pods

This is a fast and easy way to use fresh peppers from the garden.

Ingredients
Goat cheese
Small sweet pepper pods
Salt and pepper

Wash sweet peppers, slice in half, and remove the seeds. Fill each pepper with crumbled goat cheese, then top with salt and pepper. Refrigerate until ready to serve.

Fried Trout

Billy's all-time favorite fish is trout, with the heads attached. Because he does not keep many groceries on hand, I prepare these at home, wrap them in aluminum foil, and call to make certain he is ready for lunch before I arrive. Billy eats the leftovers cold for breakfast.

Ingredients

4 large trout, cleaned with heads and tails attached

1 egg

½ cup milk

¼ cup plain, all-purpose flour

½ cup cornmeal

1 teaspoon garlic salt

1 teaspoon ground pepper

3 tablespoons vegetable oil (or enough to cover the bottom of a large cast-iron skillet)

Break egg into bowl and add milk. Mix well.

In plastic bag, add flour, cornmeal, garlic salt, and pepper. Tip: recycled bread bags work great. They are long and skinny. A perfect fit for the trout.

Dip trout into egg batter. Place into bag, close bag, and shake well until fish are covered in cornmeal.

Place skillet on stove. Add oil, and heat until it is almost smoking. Add trout. Cook until flesh begins to separate from the bone, and it flakes when a fork is inserted into flesh. Turn trout over. Trout can also be baked in a 400-degree oven until tender and flaky.

8

Saving Seeds

The task of organizing Billy's deep freezer was an all-day affair that required the separation of beans classified as food from identical-looking beans he plants each spring. In order to prevent weevils from eating the seeds, Billy stores them in the freezer. The job demanded a label maker, enough storage containers to hold approximately 2 cubic feet of dried seeds, and my ability to complete the project before Billy figured out what I was doing and decided to help.

He had previously extended an open-door policy and encouraged me to implement garden improvements as I saw fit. One look at the bags, sacks, kernels, and unidentifiable items inside his freezer, and I knew this task needed what he called "a woman's touch."

The adjustment from caretaker to widower hasn't changed Billy's waste-not lifestyle. He uses items until they are past gone. For example, once a bath towel shows signs of wear, he cuts it into small pieces and uses the remnants as washcloths. Later, when the washcloths become threadbare, he uses them to dry tomatoes and other vegetables.

"I'm not a housekeeper," he has told me more than once. "My friends ask me why I don't find a woman to take care of me." Billy laughed and gestured around the cluttered living room, then quickly added, "This ain't no place for a woman. She couldn't stand living with me."

Billy isn't exactly the kind of person who needs to be taken care of. His daughters have hired someone to knock the dirt out of the house. For the time being, that arrangement works. However, there wasn't enough money in the world to hire someone to tackle the plunder building. Only I was so brave, or foolish.

Billy chose this particular title because the outdoor structure is literally stacked floor to ceiling with boxes, bags, tools, animal feed, and a variety of other plunder that only God and Billy Albertson could identify. Nestled in the back of the building, an antiquated Sears and Roebuck chest freezer served both as food and seed storage. Crammed inside were multiple paper bags that Billy had rolled down, rolled up, opened, closed, and crumpled so many times that the dirt and oil from his hands blended with the pulp, giving the bags a cloth-like feel. Inside the large sacks were smaller sacks, either unmarked or mis-marked. It is common to find bags marked "corn" that, when opened, contain beans. Ears of corn, husks intact, are jammed into every corner of the freezer. The rest of the plunder building is similar in decor. Piles. Stacks. Haphazard cast-offs. Half-empty paint buckets. Boxes piled head high. Mice. You name it, the plunder building has it…somewhere.

Lord, help us all. I cannot live like this.

In my mind, I thought that if I started cleaning the plunder building, beginning with the freezer, I might be brave enough to tackle the rest of the area one section at a time. After covering an outdoor table with a sheet and arranging empty containers according to size, I parked a wheelbarrow at the freezer and filled it with miscellaneous sacks. Two minutes into the project, Billy arrived. I had hoped to surprise him with a neatly organized, alphabetized freezer. Alas, as with most plans made at the farm, I was interrupted.

"What-chew up to?" he asked.

I gestured to the stack of recycled containers and said, "Oh, I figured I would sort the seeds today. You know, get us ready for spring planting."

Billy shrugged off his jacket, a sign I interpreted as a hot flash. He and I seemed to alternate this affliction: his brought on by hormone injections, mine from lack of hormones. Yin and Yang. Frick and Frack. Abbott and Costello.

"Some of these seeds are old and need throwing out," he said while peering into a clear plastic bag. "Let me go get a bucket."

Incorporating a bucket into the equation meant one thing—my simple, one-hour project had morphed into an all-day one.

Waving for my daughter to come help, I whispered, "We've got to hurry this project along."

Frequent visitors to Billy's acknowledge that the farm is a time warp, where a ten-minute visit quickly becomes an all-day outing.

While Jamie labeled jars with stickers, Billy culled the seeds he said were "too old to sprout."

"The chickens will just love these," he said as pink-eyed purple hull peas pinged against the plastic bucket.

I didn't want to argue, but I thought seeds could stay frozen for years, decades even, and still germinate under the right conditions. What if we needed the seeds later? My plan was to organize, not throw away. This little organizing project had spiraled out of control. Needing a diversion, I thrust a crinkled brown paper bag into his hands and asked, "Are these the beans you saved last year, or are these for eating?"

To me, they looked identical. None of the bags contained a date. How could Billy know he wasn't tossing viable seeds from last year?

The previous season, and for seasons as far back as he can remember, Billy encouraged beans to go to seed. This means that he allowed the bottom fruit to grow fat, turn yellow, and eventually dry on the vine. The process drove me nuts. I was born a gatherer, a harvester, a no-bean-left-behind kind of gal. I pluck everything from the vines even when instructed otherwise. I remove the dried pods and hide them at my house. In the fall, I return them to Billy's freezer, efficiently labeled and ready to plant come spring.

The previous year, Billy had instructed, "Zippy leave those yeller beans on the vine. They'll give me something to do this winter."

Billy, who always had some kind of project occupying his time, preferred to bring the vines indoors. He collected the wrinkled pods and then separated hull from seed. The result was a house full of desiccated leaves, dirt clods, and unhappy daughters who took one look at the floor and shook their heads. After collecting all the beans, Billy tossed them into an unmarked paper sack, rolled the top down a couple of turns, and crammed the bag into the freezer, forgotten until he gave the order, "Zippy, it's planting time. Go fetch me some seeds."

And because all unmarked paper sacks looked the same, I ended up bringing half the freezer content for Billy's proper identification.

Preserving corn seed was a different process. After harvesting a few ears, he tossed them into a bucket. Loosely protected by the husk, kernels withered and dried on the cob. If the winter was long and Billy got bored, he removed the kernels with his thumb and used the cob as a fire starter. Most of the time, he chunked the entire cob, husk and all, into the freezer, which was the reason I started "operation seed organization" in the first place. Billy knows how to grow things, but I specialize in organizing. One glance at the mayhem inside his freezer, and I thought I had died and gone to heaven. Dormant seeds waited for the opportunity to grow and thrive. They only needed a little tender loving care and perhaps a woman's touch.

❧

By January, seed catalogs arrive in the mail, bringing with them the ailment known as spring fever. Nursing this illness, Billy treats his symptoms by perusing Grier's Almanac. Tried, true, and Billy Albertson-approved for over fifty years, this particular almanac is Billy's publication of choice and free to all who enter the Wender & Roberts Pharmacy in Roswell, Georgia. It is a barebones, newsprint document with clear instructions and exact planting dates. In other words, it is fool proof. Customers have brought Billy all sorts of almanacs, which he graciously accepts, but at the end of the day Grier's Almanac is our go-to manual.

For me, shopping is the only cure for spring fever, which is why the Botanical Interests Seed Catalog, also free, is a dangerous publication. It arrives in the mail at the exact time when I need a diversion. Reaching for a pad of sticky notes, I leaf through the catalog and feed the fever.

"Oh, I've just gotta try this," I say while my husband shakes his head. Soon the pages are crinkled and adorned with multicolored markings. Later, when a brown box arrives bearing my name, I cannot wait to touch the packets. My family thinks I am obsessive-compulsive, but the sound of the seeds shaking in paper packets makes me smile. Inside are tiny pearls of promise and hope.

After converting shoeboxes into efficient sanctuaries, I separate the packets by category. There is a section for seeds collected from treasured friends, a stack of those given to me by readers, and fancy purchased seeds. Satisfied that they're all in their places, I align the packets and set them aside until spring.

ꝏ

By definition, the word "heirloom" means an object passed down for generations. I believe heirloom seeds should be shared, not purchased. Two of my favorite varieties are Floyd's Precious Watermelons and Papaw's Prize Pumpkins, commonly referred to by friends and family as "The Triple P." Neither are available in stores.

Floyd Waldroup was a mountain man. Like many men born in rural Graham County, North Carolina, he bonded with the mountains in a way that time and technology have now forgotten. Often walking the ridgelines with his children, and later his grandchildren, he identified plants, shared their usefulness, and every now and then dug a sprig or two of ginseng. He owed the mountains a lot because they led him to Lena, the love of his life. Floyd was just a mere sapling, clearing the land behind Lena's Paw's place, when he noticed her working outside. At that moment, he became so thirsty that he had to stop and ask for a drink of water. Lena was seventeen when she married Floyd. Their marriage lasted sixty years.

Floyd was small in stature, big in heart. After a long day of clearing property and skidding trees, he found comfort in his garden. He walked the rows several times a day just to watch stuff grow. There were no weeds in Floyd's garden. His repetitive steps pressed the dirt flat and warned weeds that they dare not enter. When I think about the legacy of Floyd's Precious Watermelons, I smile. I like to imagine that Floyd looks down on us with his kind blue eyes and smiles too.

I obtained the watermelon seeds that bear his name from Floyd's daughter, Linda, who acquired them in a most unusual manner. One summer, Floyd enjoyed a refreshing slice of melon on Linda's front porch. After wiping his mouth to remove sticky pink juice, he rocked back as he chewed, then forward as he spat the seeds into the front yard where they landed at the base of a butterfly bush in an area thick with mulch. The following spring, while pulling weeds, Linda noticed the vine now called Floyd's Precious Watermelon. She could have yanked the wayward vine from her flowerbed and tossed it in the sun to teach the uninvited guest a lesson, but she did not. Neither knew that cancer would take Floyd a few months later.

Nurturing this plant, Linda coddled, cried, and coaxed a single watermelon. One melon was all it took the first year to provide what is now the heritage of Floyd Waldroup. Those small black seeds magically connect Floyd to his family. They are a treasure. A gift. A memory that lingers in the heart eternal. We proudly grow them on Billy's farm.

Seeds planted in your heart take root and bloom forever.

ꝏ

I can only imagine where my grandpa procured his pumpkin seeds.

John A. Parris was one of many people who sat for a spell at Winchester's Grocery. His featured *Asheville Citizen (Times)* column, "Roaming the Mountains," introduced readers to my grandpa and my great-grandpa, Columbus "Lum" Winchester. Parris visited Winchester's Grocery often, hoping to glean wisdom and a good story from Lum. This is how I suspect the seeds found Grandpa—through either a casual visitor or a regular customer. It is also possible that Grandpa acquired the seeds from a Native American, especially since the Cherokee Indian Reservation is nearby, and a portrait of his grandmother, "Aunt Winchester," once hung in the Great Smoky Mountains National Park Visitor Center at Oconaluftee for the public to enjoy. Perhaps Aunt Winchester planted the seeds as livestock fodder and passed them to her son, who shared them with my grandpa. No one knows for certain. Regardless of lineage, I am blessed to possess seeds that produce The Triple P.

Before an image of *The Great Pumpkin* forms, I should explain that Grandpa grew the Cushaw variety, not carving pumpkins. Despite hours of Internet searching, I have yet to find images of the variety he grew. For that reason, I share the seeds only with people whom I know for certain will treasure my gift. The Triple P is a round variety of Cushaw with an outer watermelon-colored skin that changes color several times during maturation. Freckled leaves the size of elephant ears protect swelling fruit from the sun. As summer fades to fall, the leaves gradually die back, revealing a pale yellow-orange pumpkin that weighs between 10 and 30 pounds. When baked, these pumpkins make delicious pies.

Old-timers with years of farming experience recognize the Cushaw for its beneficial nutrients. Some still grow it for livestock feed.

In the Winchester family, Grandma Wonderful possessed the business know-how. Only Warren Buffett was wiser, though not by much. After properly stocking her barn, she displayed surplus pumpkins around a maple tree at the edge of the driveway. Visitors who stopped while Grandpa was there could talk him into lowering the price. No so with Grandma Wonderful. After raising seven kids, eight if you count her husband, she knew her crop was superior. Hers were a rare variety grown by no other farmer in the area. Memorizing grocery store prices and the market value of produce from as far away as Asheville, Grandma Wonderful knew their value. Trying to negotiate her price was an insult.

That'll be five dollars apiece please, ten if you haggle.

While some may define heirloom seeds as a pure variety without genetic modification, for me the connotation triggers a memory and a personal connection that sometimes brings a bittersweet tear and always brings a smile.

ᘓ

Billy obviously does not share my emotional attachment to seeds. I had no idea that he would toss a single kernel. As the chickens gorged, I wondered, *Where are all the visitors who usually drop by?* Any other day, carloads of customers clogged the driveway. I sorely needed a visit from the legions of folk whom we expected to dropped in. As I silently hoped that a Chatty Kathy would appear before Billy emptied everything into the chicken lot, a flash of blue caught my attention and brought the entire project to a halt.

"What are those?" I asked while quickly reaching for a pimento cheese container Billy held.

"I dunno," he said as he ripped off the lid. "Something someone gave me. People are always giving me stuff hoping I'll plant it for 'em."

He placed a blue seed in his palm and rolled it around with his finger. After a thorough examination, he said, "Can't say I know what this is," and tilted the container toward the chicken bucket.

Grabbing the plastic cup, I said, "Let me have a look." I poured a few kernels into my hand. Deep indigo in color and rock hard when pressed, the seeds resembled sunflower hearts, only fatter and pear shaped. Tucking them inside my pocket, I announced, "These are too pretty to feed to the chickens. Surely they will grow something beautiful."

Billy shrugged. A gesture I took as a sign to mean they are all yours.

"I'll take them home and do a bit of research. I'm sure someone knows what these are."

Moments after posting images on the web, I found my inbox filled with messages. The indigo-colored seeds were cotton, one of the most delightful crops in the South.

Pumpkin Pie

Ingredients
1 unbaked pie shell
1 small pumpkin
1 tablespoon cornstarch
1 egg
1 cup brown sugar
1 cup evaporated milk
1 teaspoon cinnamon*
½ teaspoon allspice*
½ teaspoon ginger*
½ teaspoon ground cloves*
½ teaspoon nutmeg*

*Note: If desired, replace the above spices with 2 teaspoons of pumpkin pie spice found in your grocery store.

Preheat oven to 350 degrees. Slice pumpkin in quarters. Remove seeds and reserve for toasting (recipe follows). Place pumpkin facedown (rind up) in a deep cake pan and add approximately ½ cup of water. Bake until tender, 45-90 minutes. Bake times vary depending on size and weight. The pumpkin is ready when the rind darkens and the flesh is soft and tender.

While pumpkin cools, use a blender or food processor to mix brown sugar, cornstarch, egg, spices, and evaporated milk. Mash pumpkin with a fork. Discard rind. Add to spices. Process until smooth. Spoon into unbaked pie pan. Bake for 45-50 minutes.

Toasted Pumpkin Seeds

While this recipe calls for two cups of seeds, not all pumpkins contain exactly two cups. As Grandma Wonderful says, "Use what you've got to make what you need."

Ingredients
2 cups pumpkin seeds
Salt
1 tablespoon butter
Nonstick spray

Preheat oven to 250 degrees.

Melt 1 tablespoon of butter and pour it on a cookie sheet. Arrange the seeds in a single layer. Lightly salt and then spray seeds with nonstick spray to bond the salt to the seeds.

Bake for 45 minutes to an hour, or until lightly brown. Occasionally toss seeds to ensure browning on both sides. Enjoy immediately.

Fried Egg Sandwich

The tradition of this sandwich goes way back. Each Sunday after church, Grandma Wonderful's kitchen fills with kids, grandkids, and great-grandkids. When I was growing up, the eggs were fresh, gathered from her henhouse located across the road, but she stopped raising chickens years ago. Today the eggs are store bought. Grandma Wonderful stands at the stove, electric element turned to high, skillet sizzling with eggs. She calls out, "Who all wants their yellow busted?" The yesses receive a sandwich with solid egg; others prefer the pleasure of a gooey yellow middle when they bite into their sandwich.

When my daughter's friends come to our house, I offer them the same option. There's just something magical about this sandwich. It is my family's Sunday tradition.

Ingredients
2 slices white bread
Butter
Salt and pepper to taste
1 egg
Duke's mayonnaise (if desired)

Butter the bread and place in oven at 350 degrees to toast. This is a key step. You want an evenly toasted slice of bread. Toasting on broil will most likely brown the edges first, making a hard crust. Of course, you can decorate the toast like Grandma Wonderful with dots of butter for eyes, nose, and a mouth, but I digress.

While bread is toasting, place a dollop of butter in the skillet. Crack egg. Add a pinch of salt and a shake of pepper. Egg is ready to turn over when you gently tilt the skillet and the white doesn't run. If the order is for a "busted" yellow, pierce the yolk and allow it to cook, then carefully flip the egg.

Assemble on toast and enjoy.

9

Cotton: From King to Contraband

I can't rationalize why I love cotton. Perhaps it is the blinding white bolls presented on prickly rust-colored limbs, a combination of hard holding soft, brittle displaying beauty and strength. Maybe it is the community that was once necessary to plant, cultivate, and harvest the crop that lures me to its magical spell. My people didn't grow cotton. I grew up in tobacco and cattle country. Cotton is not part of my heritage. Even so, every time I see a field of cotton, something inside my soul longs to touch the fluffy bolls.

In the Georgia flatland, cotton fields can steal your breath. A driver has no choice but to slam on the brakes and ignore the passengers protesting. Stepping into the field, I marvel at how the seemingly impenetrable clay produces a river of white. Standing on tiptoes while shielding my eyes, I gaze across a field as the bolls continue their peaceful journey, rolling as far as I can see. In the distance, the sky dips down, blue touching white, heaven kissing earth in a glorious display.

Snow in the South is a beautiful and magical thing—especially when it's made of cotton.

Seeds, small and insignificant, have pierced the rock-hard spring soil. The plant has fought element and insect, presenting to the world a sea of purple and white blooms with one goal: the production of soft fibers. Land, divided only by blue sky and the occasional cumulus cloud, provides a stunning canvas. Metropolitan travelers often drive through Georgia's cotton country en route to tropical destinations. Many ignore this miracle, this unpretentious fiber, not realizing or caring that the lowly cotton boll permeates virtually every life on this planet. Cotton is much more than fabric and cloth; it is a multimillion-dollar industry. I may call the plant beautiful, but in flatlands across Dixie, folk call it "White Gold."

Billy Albertson remembers picking cotton.

"Poppa was a sharecropper until I turned thirteen," he told me. "By then I was old enough to help other farmers. Ed Wood lived on Freemanville Road. He always grew cotton like most folk did back then. Since Momma and Poppa had so many of us kids, Old Man Wood asked Poppa if I could help pick. He paid me a penny a pound to pick his cotton. You know that was a lot of money back in them days. I carried a pick sack into the field, and before the day was over I had picked one hundred twenty-five pounds."

The moment friends identified the indigo-colored seeds I'd found in Billy's pimento cheese container, I snatched the phone off the cradle and dialed his number.

"Guess what we're planting?" I asked.

Knowing me well, he responded, "Ain't no telling."

"Remember those blue seeds we almost threw out? They're cotton."

"Cotton? Well, well. Ya don't say. I reckon you'd better bring them back over here. We gotta get those seeds in the ground."

Now it was time for the bad news.

"First, we need a permit from the Georgia Department of Agriculture."

"I need a permit to grow cotton?"

"Yes. It's illegal to grow cotton without a permit."

"Good night!" Billy said. "It's illegal to grow cotton, in Georgia of all places. Well, I've never heard of such."

Those were my thoughts exactly.

Billy continued, "You mean to say they got somebody down there in the state capitol that ain't got nothing better to do than worry about me growing a handful of seeds?"

I responded by reading information from the *Overview of Boll Weevil Eradication Program*. "The Director of Agriculture sent me a copy of the regulations. According to it, I need to measure the area, give them directions to your place, and state our intent for growing cotton."

"Our intent?" Billy said. "Well, our intent is to grow a row of cotton so folk 'round here know what it looks like. We ain't gettin' in the cotton growin' business, if that's what they want to know."

He paused. I knew what was coming next.

"How much does this permit cost?"

Thankfully, the paperwork only cost my time. Anything else was highway robbery in Billy's book. Mine too.

I explained. "Since we're a private organization, there is no fee. I figure if I write the director a nice letter explaining how many home-schooled children come to the farm each year for field trips, we'll get a permit."

Satisfied, Billy said, "Send it out. Kids around here need to know about cotton."

❧

As the permit traveled through the chain of command, Billy and I surveyed his property for an adequate spot.

"Cotton takes a lot out of the soil," Billy said while approaching the plunder building. "What do ya think about putting it here?" He kicked the soil with his boot. "Won't nothing much grow here 'cept weeds."

The selected location was more gravel than dirt. Glancing around for other options, I said, "I think I'll call the director and leave him a message to see how much longer before we get approval."

Walking the tiller to the area he'd chosen, Billy said, "I'll just till us up a little spot while we wait on our permit."

"What if they say no?" I asked.

"We'll plant it anyway."

Section 40-24-1-.13 of the Georgia Rules of the Department of Agriculture states, "Non-commercial cotton shall not be planted without a permit." Section 40-23-1-.17 lists the penalty for growing ornamental cotton without a permit as "a fine of not less than $50.00, nor more than $1,000, or by imprisonment not exceeding 12 months, or both."

"I'm not worried," Billy said after I explained the risk. "We've only got a handful of seeds. Surely they won't throw an old man like me in jail over a dozen stalks of cotton."

I didn't tell Billy about the town in middle Georgia that had planted cotton in an effort to celebrate their heritage and beautify downtown. That little mistake cost them a cool thousand after someone failed to obtain a permit. I was nervous and silently prayed for approval as we waited. Spring arrived. Vegetables sprouted without word from the state.

"Heard anything from the Department of Agriculture?" I asked, just in case Billy had received the permit instead of me.

"Not a peep."

I resubmitted the application, the letter, and the rendering—this time by e-mail. We waited some more.

"It's time to plant the cotton," Billy announced one day.

Excited, I asked, "Did you hear from the director?"

"Naw. But I figure it's now or never." Passing the seeds to me, he said, "Let's get started. If farmers waited around on the government, folk would starve to death."

After reminding him of the financial risks, not to mention jail time, I followed behind and carefully placed indigo seeds into the earth. The spot he selected wasn't tucked behind the shed. No sir, he tilled up a section beside the house, almost in the front yard. The area was in plain view, right out there in front of God and everyone else.

Since the paranoids were after me, I sprinkled Sevin Dust the moment tender leaves emerged from the earth. Weevils hadn't been reported since 2002. No one, especially me, wanted to hang out the welcome sign for that particularly destructive pest.

In 1915, *Anthonomus grandis Boheman,* commonly known as the boll weevil, began destroying the cotton crop in Thomasville, Georgia. By the 1960s, farmers were using pesticides to eradicate the creature. In the 1980s, however, researchers discovered that pheromone traps worked as effectively as poison. Cotton farmers, whose livelihood hangs at the mercy of this creature, obviously needed to prevent another outbreak at all costs.

It only takes one infestation to wipe out a multimillion-dollar industry. This is why Billy and I searched each plant every single day for weevils.

As expected, folk noticed our crop. Many customers pulled into the driveway and asked, "Is that cotton?" Billy just smiled and guided them to the rows, where he described the anatomy of the plant. Cotton contains three parts: lint, hull, and the kernel. The lint part makes yarn, cloth, and other products we use. The short fuzz attached to the seed is used to produce plastic, cosmetics, upholstery, and paper. Hulls contain a high percentage of protein, making them excellent livestock feed. Crushed kernels produce cottonseed oil and meal for human consumption.

The dozen seeds we had planted wouldn't produce enough to sew a sock or feed a gnat, but they did provide an educational opportunity for children and adults alike. We explained to visitors that cotton is contraband and asked them to refrain from snapping photos and posting them on Facebook. They agreed and shared their personal stories instead.

One visitor described an arrangement between the local school system and farmers desperate to harvest before winter rains came. "In order to be eligible for the senior trip to Washington, DC, our class had to pick cotton. Since none of us had ever been out of the county, much less the state, we all volunteered. We were excused from school and spent all day in the field."

Her friend said, "The way I remember it, students didn't have much of a choice. If you wanted to go on the trip, you hit the field."

I like this concept of farmers partnering with schools. Picking cotton provided those students with an excellent way to experience—if only for a small moment of time—the back-breaking, finger-piercing work that generations of slaves had done before them. Their community service work served as a living history lesson. For that small moment, the students worked the same field the slaves had worked—those slaves who had draped a pick sack across their skinny shoulders and started the day fueled by a breakfast of sweet potatoes, hoecakes, and sausage (if they were lucky). They picked from daylight till dark not because they wanted to go the nation's capital on a high school fieldtrip, but because they had no choice but to harvest snow-white bolls of cotton as the Georgia sun beat down on fields of clay.

There was a time when students started classes after Labor Day, and schools closed when it was cotton-picking time. And the job wasn't only for slaves. If you could walk, bend over, and pinch your fingers together, you picked. Pause for a moment and consider doing this with the youth of today. Can you hear the cries of abuse, neglect, lawsuits, and refusal to work?

"Kids these days wouldn't strike a lick if their lives depended on it," Billy said. "Back then, if you didn't pick, you didn't eat."

Stories swirled around the garden. One woman said, "I know someone stole from my pick sack. I worked all day long, and at the end of the day my sack didn't weigh enough for me to earn a trip." Another lady added with a knowing smile, "I worked with my boyfriend. He picked enough for both of us."

Those were the days when cotton was king and textile manufacturing the crown jewel. When textile manufacturing moved from the United States into other countries, the fabric of our livelihood slowly unraveled. Oh, how my heart longs for the return of towns humming with machines that weave hope into our lives.

Billy isn't the only person with farming in his DNA. Legions of modern-day folk descended from dirt farmers. They were hardworking entrepreneurs whose sweat-tinged brows dripped beads of hope into the soil. This was before urban sprawl spread through the country like kudzu. If you have ever traveled through Georgia, odds are you have driven over land that was once cotton country. Beneath all that black-topped highway is land that once grew crops. Long before peanuts and soybeans, cotton was the pride of the South—a significant plant weaving the cords of heritage and history, whether we choose to acknowledge or ignore it.

Believing that cotton growers held an unbreakable power, on March 4, 1858, South Carolina's James Henry Hammond proclaimed, "Cotton is King." While the beautiful plant no longer holds that title, it still plays an important part in our lives.

I should clear up any misunderstanding that might arise from my cotton-growing confession. Billy and I had an insect-eradication plan. Of course, we realize that the boll weevil brings a serious threat. We are responsible farmers. We monitored the crop on our hands and knees. From the second the first green blade erupted until the bolls offered their harvest, we analyzed, observed, fretted, and grew proud of our plants. Still, as we celebrated each stage of growth, Billy and I couldn't understand how a simple farmer needs a permit to grow ornamental cotton in the Land of the Free. This type of regulation must have James Henry

Hammond rolling in his grave. Mr. Hammond famously said, "No, you dare not make war on cotton. No power on earth dares to make war upon it." I was shocked that a veteran farmer like Billy could receive a fine and jail time just for growing an innocent patch of cotton on his own property.

More complicated issues also puzzled us. Why are textile mills empty as Americans search for jobs? Why do companies genetically modify crops? Why does this country import food, lumber, and cotton? What happened to the days when workers picked, ginned, milled, and sewed cotton all in the same town?

❧

Cotton is a distant cousin to the hibiscus flower and the delicious okra vegetable. The stalks display stunning blooms that are either white or purple, depending on the stage of the life cycle.

As our crop grew, each trip I took to Billy's began with a stroll through the plants, touching each one and documenting the progress with photographs of every stage.

"I'm right smart proud of our little stand of cotton," Billy said.

I was too. As the plants grew, leaves unfurled and reached toward the sky. The short-lived blooms gave way to flat, triangular pods that hid beneath the foliage and eventually changed from green to purple. Many months later, a ball of fluffy cotton opened. We had waited for this day.

"Ya know, this whole place used to be white with cotton," Billy said, looking around the property. "It used to be as far as the eye could see a river of white. Everyone 'round here grew cotton. Every town had their own gin. The boll weevil killed all of that. After the boll weevil, folk started raising chickens. Now that's gone too. All folk have now is big houses and important jobs that keep them busy."

That is why Billy and I planted a little trickle of white in the very same spot where a mighty river had once flowed. This was our cotton, Billy's heritage. It may sound odd, but I didn't want to pick the cotton. I was in love with the plant, so comforted by its welcome at the entrance of the farm that Billy had to force me to harvest.

"Rain's a coming," Billy announced. "Let's get the cotton inside before it ruins."

I was going to miss the limbs that had faded from a royal maroon to a dull brown. "This little patch of soil might not grow veggies," I said, "but it gave us something beautiful this year."

"Make sure you save some seeds," Billy said. "They're inside the boll."

Rumor has it, and Hollywood portrays it as truth, that the process of picking cotton is an arduous task. One filled with sweat, blood, and sometimes tears.

"Now, what you want to do is avoid the dried prickly part," Billy advised while watching over my shoulder. "Just reach down and pull the white boll from the plant."

Using my thumb and forefinger, I removed the cotton in a slow, dainty method. Wanting to gauge Billy's response, I jokingly said, "I don't know what all the fuss is about. This cotton pickin' is a piece of cake."

Billy threw back his head and laughed.

"That's because you've got all day to pick these few stalks. You're not dragging a sack behind you while Old Man Wood waits for you to finish."

I laughed too.

"You're also standing in the shade. Ain't no shade in a mile-long field of real cotton."

We laughed again, a knee-slapping moment.

Cotton is nature's paintbrush. It is a masterpiece of hope, a story of survival against all odds. It is a celebration of life in its purest form, a moment of heritage growing in the Georgia clay. Our little patch of cotton provided me and others a glimpse of Billy's past and created newly made memories.

Poor Richard's Cotton Candy Recipe

For grandparents who volunteer to take children to the cotton festival, this recipe will empty your pockets speedy quick.

Directions

Drive to cotton festival or county fair. Purchase admission for yourself and all the children. Remove $5.00 from pocket. Give to cotton candy vendor. Repeat as often as necessary to silence whiney grandchildren.

6 months later: Pretend you have no idea why your grandchildren have so many cavities.

Pimento Cheese

Why buy pimento cheese when you can whip up a batch in three minutes? The secret to tasty pimento cheese is a drop of prepared mustard.

Ingredients
8 ounces shredded cheddar cheese (may use sharp if so inclined)
1 small jar pimentos (drained)
1 heaping tablespoon Duke's mayonnaise
Prepared mustard
Fresh cracked pepper

Mix cheese, pimentos, mayo, and a drop or two of prepared mustard together in a small bowl. Sprinkle with fresh cracked pepper.

Serve with celery or as a sandwich between two slices of bread.

10

Little Ones in the Garden

Feet bare. Shirtless. Sweaty. Skin tickling and itching as salty drops of moisture trickle down his back. Billy Albertson, age knee-high to a mule, works a piece of property he will never own. Egbert Albertson, Billy's father, whom he calls "Poppa," works alongside him. Poppa is a sharecropper and will be for several more years. It is a time when the United States is recovering from the Great Depression. Sharecroppers live on and work land that belongs to someone else. In a sharecropping arrangement, crops serve as currency. Large families are a part of this arrangement. Each member performs a chore such as milking, planting, or chopping wood.

"Back then, if you didn't work, you didn't eat," Billy explains when asked to reminisce.

Schoolchildren often visit Billy's farm today, especially those with class projects on the Great Depression. Billy explains that historic time with accuracy.

His connection to the land began early. As the baby of the family, he toiled beside his siblings until they married and started their own families. When asked how many siblings he had growing up, he offers a confusing reply: "I grew up in a small house on Etris Road. The family had seven boys, and each of those boys had four sisters."

To clarify, there were four girls and seven boys. Each boy dearly loved his sister. The feeling was mutual. Slowly, members of the family unit married and started their own lives until there were only three in the home: Momma, Poppa, and Billy, who believed it was his responsibility to care for his parents.

"Helping Momma and Poppa just came natural," Billy says. "I felt like it was my job to help them out."

He recalls the handmade clothes. "Momma stitched them from feed and flour sacks. I milked the cow for Momma every single day. I still remember the last time I milked for her." Billy looks out the window with an expression of longing for a time long passed. "We had one pair of work overalls and a Sunday-go-to-meeting outfit. Folk had only one or perhaps two pairs of shoes. Most of the time we just ran around barefoot. Yup. The last day I milked for Momma was the day I got married."

Life was slower and quieter in the 1940s. Devoid of concerns about health issues like tetanus and staph infections, kids stayed outside all day. They worked and played in bare feet. Perhaps this is why Billy ignores the dangers rusty nails bring and dashes into the garden barefoot. Perhaps feeling dirt on his bare feet today reminds him of home.

ᴥ

When children visit the garden, Billy experiences pure joy. Bending to their level, he greets visitors with a smile and says, "Looks like you need a farmer's hat."

With that, he plunks a worn straw hat that is ten sizes too big atop the tiny future gardener's head. Children of every age are welcome at Billy's. Those who are well behaved might even have a chance to feed the chickens or the goats, especially in the winter when he isn't too rushed with other responsibilities. Spring visitors who arrive early in the morning have the opportunity to get their hands dirty.

On more than one occasion, youngsters have said, "Mr. Billy, do know that you can get all of these vegetables at the grocery store? That way you won't have to work so hard."

This prompts him to say to the parents, "Kids these days need to learn about food. They need to understand that food doesn't come from the grocery store."

Billy is a self-appointed soil ambassador. If you aren't dirty when you leave his place, then he's not happy. Glancing toward their parents for approval, the children touch the soil with their tiny hands, perhaps for the first time. Sweat replaces uncertainty. Dirty fingernails become badges of honor, a reward for a job well done. Perhaps that is the best part about touching the earth: it teaches patience. Many schools now have small raised-bed gardens. We are blessed when they invite us to their outdoor classrooms. This is the perfect environment for Billy and me. We love teaching the youth about seeds, about planting and growing their own food. With each seed, each open flower, and each ripe vegetable, uncertainty and self-doubt disappear, and confidence grows.

ᴓ

Wearing trendy rubber Croc shoes, designer overalls with matching caps, and layers of sunscreen, today's young gardener does not resemble a young Billy Albertson. Even with physical differences, though, similarities remain. There is a seed placed inside a child's heart during his or her creation. When nurtured, that seed helps the child grow into an adult who respects and treasures the earth.

Adults merely need to nurture a child's curiosity. Teach kids that getting dirty is a lot of fun. Place a shovel or spade in a child's hand, and watch her get to work. Dirt flies through the air. Holes appear in the earth. Smiles adorn her face. The simple task of growing flowers and vegetables can increase self-esteem and improve confidence. Where else can one take an insignificant seed, cover it with dirt, and watch something magical happen? Children need to create; to grow as individuals; to do something that maybe even their parents cannot do. In order to respect the earth, our youth must first understand that it belongs to them. Even if their attempt fails and parents must strategically place a flawless grocery store cucumber beneath the leaves of a pathetic plant, kids need to touch the soil.

Knowing firsthand the importance of getting dirty, Billy and other farmers like him welcome everyone into their garden. Understandably, in today's world, parents should use caution around strangers. I am not suggesting dropping children off unsupervised. Instead, consider befriending someone much older, even if you don't garden. The memories created will last a lifetime, and the person who benefits might just be you.

For many, this suggestion is difficult to grasp. Our youth-celebrating society prefers to tuck senior citizens out of sight. But when we do this, we overlook the obvious: we all age, we all grow older, and we all, God forbid, might find ourselves alone in the blink of an eye.

ꝏ

Neisha Handley is the mother of Zoe, Sofie, Jairus, and Mira. After reading the first book about Billy, she decided to step out of her comfort zone and find him. She pointed her car toward Roswell. What she found there was more precious than fresh produce.

"I was returning from Trader Joe's and noticed a sign that read 'Hardscrabble Road.' I told my husband, 'There may be seventy-one Peachtree roads around here, but I bet there is only one Hardscrabble.'"

By the way, according to Billy, the people of Hardscrabble Road named it that because "times were hard and people had to scrabble enough money just to make ends meet."

When I asked Neisha about her first Billy experience, she said, "After I pulled up, Billy stepped outside and said, 'Well lookie here.' He acted as if he was expecting us. He had an instant smile. There were hugs and handshakes. Instantly, the children were enamored with him. Then he said those magical words, 'Come on back here and have a look around.'"

Neisha grew up on 86 acres in Missouri. Like many others, she and her husband live where the jobs are. She is far from home. Homesick. Atlanta's concrete jungle has her trapped, desperately searching for something to connect her to memories of home. Big corporate jobs, while providing a better way of life for our children, separate us from parents and grandparents. Our people wanted this better life for us. We left the nest and flew far from home. As time passes, we see less and less of our families, until some only visit their relatives a few times a year. Technology allows us to connect often, but nothing compares with a hug full of love. This "better" life comes with the unexpected downside of loneliness.

As Neisha's children played with the animals, her youngest child grew timid. All of that running, scampering, and baby goat adoration over-stimulated her tiny body. Instead of running to her mother for comfort, she wrapped her arms tightly around Billy's leg and said, "Will you hold me?"

Some children just know good people when they see them.

Neisha's oldest child, the adventuresome one, said, "This is where I've always belonged."

Some children know their purpose at an early age.

Billy gave them the tour. He pointed out his late wife Marjorie's blue hydrangea. He explained the difference between the raspberry and blueberry bushes. He identified fig trees. He gestured to the beans and said, "I always plant my green beans on Saint Patrick's Day, just like my brother did."

Placing a basket in each of the children's hands, he said, "Follow me," as they entered the cucumber patch.

"Do you like cucumbers?" Billy asked. "How about maders? You like tow-maders?'

Zoe, Sofie, Jairus, and Mira flashed wide smiles and nodded.

"Then let's get to picking."

When Billy extended the invitation, Neisha's children became outside children on an adventure with an outdoor man. They searched vines. They plucked cucumbers hidden beneath the leaves.

Pointing to a thick cluster of tomato plants, Billy explained, "These are volunteer tomatoes. I did not plant them. The seeds come up when the mother plant drops a rotten tomato."

The Handley children scooped up a handful of seedlings, transplanted them into a used milk jug, and renamed them "Mr. Billy Tomatoes."

They collected more than fresh veggies; the children translated the tour into useful information. The next time they went with their mother to the grocery store, they remembered where food comes from.

"Right away, you want to help Billy," Neisha said. "I can't really explain it. We no longer think about our heritage. People no longer leave a legacy. We capture photos on our phone instead of film. I want to expose my children to the upbringing that I had. Billy allows me to give them a glimpse into the life I remember."

ଷ

There is a hiding place on Billy's farm. It's one so special and so hidden that three years passed before I discovered the Best Friends' Club. Ten years ago, club members met beneath the waxy leaves of a magnolia that Billy had planted fifty years earlier. Recognizing that this tree was special, Billy's granddaughter, Kristen, claimed it and formed a girls-only group. Kristen marked her territory with a sign and then held secret meetings, carefully hidden from the prying eyes of troublesome boys. The magnolia thrived. Its limbs grew thick and heavy. Massive branches bent and touched the ground, hiding the clubhouse entrance. Club members attached a rope to one of the limbs and fashioned a shoe-sized loop near the ground. The foot-swing welcomes visitors, begging them to place a foot in the loop and kick away from the earth, to swing back and forth until their arms cramp.

Cool despite the summer heat and quiet despite a busy road, the Best Friends' Club provides children—and adults—with an opportunity to step into a world of skinned knees and calloused hands. In this adventuresome world, the deep green leaves of a magnolia tree capture whispered secrets, hide us from the rest of the world, and make us feel young, regardless of age.

Only those with a high moral code of honor are eligible to join. Membership requirements are as follows: listen to Billy, be kind to others (even your brother or sister), come when your parent calls, and never, ever carve the bark or damage the tree. She was here before you. She deserves our respect.

Adults easily overlook the knee-high sign hidden under the magnolia, but the Handley kids didn't.

"It's a tree house!" Sofie said.

Billy nodded. "My granddaughter and her two friends, Katie and Amber, started this club a long time ago."

Instantly the Handley children made their membership requests known. "Mr. Billy, can we be members of the Friends Club? Can we? Can we, please?"

"Why, shore."

Neisha's children ran to the tree and disappeared behind the evergreen leaves. Soon it was time to leave. Squeezing drops of love into

Billy, they reluctantly loaded into the van with hands full of veggies and volunteer tomato plants. They bragged to their school friends that they belonged to Mr. Billy's Friends Club and were growing Mr. Billy Tomatoes. While Neisha cannot recreate her childhood experience in her subdivision, Billy's farm offers her children a glimpse of her past. Later, when Neisha's grandmother, whom she calls "The Mam," visited Atlanta, neither were interested in a trip to the Georgia Aquarium or the World of Coke. Both had a hankering to visit Billy's, where The Mam also became a member of the Friends Club.

Shakespeare said, "One touch of nature makes the whole world kin." Exposing children to the garden grows more than vegetables. It plants memories deep in their hearts. The Handley children will never forget their time with Farmer Billy. Now they are kin, just as you will be when you visit.

ꝿ

I am on a mission to rid the world of lonely people. Give a little bit of yourself to a stranger just as Neisha and her children did. Manufacture an excuse to check on the elderly and those who live alone. If you believe that everyone has a story to tell, start there. Practice the art of visiting. Ask questions. Share your life with someone else. You just might earn a membership into the Friends Club and find yourself playing beneath the magnificent bow of a magnolia tree.

Beanie Bags

Healthy eating habits and life lessons can begin inside a zip-top bag. Beanie bags are a wonderful way to encourage children to learn about nature. All you need is a zip-top sandwich bag, a napkin, and a few dried beans.

Ingredients
Plastic zip-top bags
3 dried beans (can use dried beans from your pantry)
Paper napkin
Water
Tape

Lightly moisten a paper napkin and fold it into a small square. Place the napkin in the corner of the bag. Insert a couple of dried beans near the napkin. Pinto beans are a perfect choice. Seal the bag and tape it to a window that filters several hours of sunlight. Each day, watch as the bean swells and slowly sprouts.

The best part of this experiment is getting to witness everything that normally goes on beneath the soil.

Those with limited gardening experience might be surprised to watch long roots and tiny leaves develop inside the bag. Even though there is only a small amount of water in the paper napkin, the tiny seed provides an example that adults can use as a life lesson: one never knows what is inside the heart. Once leaves develop, carefully transplant the seedling outside and continue to watch it grow.

Cucumber Sandwiches

Please do not store cucumbers in the refrigerator. Like tomatoes, they taste best when eaten at room temperature.

Ingredients
1 cucumber, peeled and sliced lengthwise, not coin shaped
2 slices of bread
1 teaspoon Duke's mayonnaise
Dash of salt
Dash of pepper (optional)

Slather mayonnaise onto bread. Place sliced pieces of cucumber on bread, and then sprinkle with salt and pepper. Enjoy.

Neisha's Granola Nibblers

This granola bar is delicious. Feel free to add or substitute the nuts and dried fruit with your favorites.

Ingredients
3 cups gluten-free oats
4 tablespoons butter
½ teaspoon salt
¼ cup Turbinado sugar
3 tablespoons molasses
2 teaspoons vanilla
1 cup finely chopped pecans, almonds, or peanuts (select one)
1 cup raisins, figs, or dried apples (select one)
1 teaspoon cinnamon
1 cup chocolate chip cookies

Mix 3 cups gluten-free oats, 2 tablespoons melted butter, and a pinch of salt. Bake at 350 degrees for 15 minutes until toasty brown. Remove from oven and let cool. Set aside for later use.

Melt 2 tablespoons butter, sugar, and molasses until barely bubbly. Turn off heat and add vanilla.

In a large mixing bowl, toss in the oats, cinnamon, dried fruit, and nuts. Pour in the melted, gooey sugar mixture and mix well. Press into a 9-x-13 pan. Return to oven and bake for 15-20 minutes at 325 degrees. Allow to cool slightly. Press again. Cut into bars.

Melt chocolate chips in microwave and use to frost bars.

11

Tools, Techniques, and Trucks

"The Chevy suits me just fine," Billy told his daughters each time they tried to talk him into buying a new truck. According to him, he had a perfectly good truck, a 1969 Chevy he bought brand-spanking new at Childer's Chevrolet in Canton, Georgia.

"I remember the day I brought home the Chevy," Billy recalls. "Janet was six years old. I told Marjorie, 'Mother, this Chevy will last me as long as I need a truck.'"

Even though metal wires hold the choke in place, and the gears lock down and sometimes require a little coaxing with a ball-peen hammer, in Billy's eyes the Chevy is as good as new. The vehicle has no power steering or air-conditioning. These options are frivolous, not standard.

As long as Good Old Reliable cranked, Billy believed they belonged together and should remain married. I agreed. Billy didn't make the decision to purchase a new truck; his daughters did. The quest for a replacement began after Billy hauled a load of wood. It is unclear whether a wayward piece of timber dinged the back window, or the rubber seal succumbed to dry rot and loosened enough to shift and then break the glass. Regardless, after Billy's youngest daughter, Denise, saw the truck, she proclaimed that "was it," and placed a call to her sister, Janet. Together, they proclaimed that the Chevy had to go. When Billy's granddaughter, Kristen, professed her adoration of the Ford F150, all hope of refurbishing the Chevy vanished.

Billy knew that replacing Good Old Reliable with a Ford could ruin his reputation and tarnish his good name. Daughters cannot understand the embarrassment and ridicule brought on by this transaction; they're unaware of the looks of pity associated with trading down from a Chevy to a Ford.

Billy and I get attached to our vehicles. It's not that we're too cheap to purchase another one, although maybe we are. It's that once you break in a ride, why go through the discomfort of learning how to drive something else? Asking Billy to become accustomed to a new road hoss was like convincing him to learn how to drive all over again. After years of driving the Chevy, reprogramming his left foot to stop feeling for the clutch would take time, as would teaching his right foot not to be so heavy. This is why Old Reliable was perfect. We just buckled up, put her in drive, and away we went. With a rattle and hum of the engine, she told Billy when she needed another gear, and he obliged. She knew the way to the places Billy frequents: the feed store, the gravel quarry, and the manure pile.

This would not be the case with the new, unnamed truck.

The battle between Chevy and Ford isn't new. Diehard Chevy owners don't just happen; they are raised that way. Most loyalists would rather push a Chevy than (gasp) drive a Fix-or-Repair-Daily, Found-on-the-Road-Dead F-O-R-D.

According to my dad, he has put a million miles on Ford pickups during his employment with the power company. Yet every single day he drove the company vehicle home and parked it a great distance away from the family Chevy.

Oil and water. Impossible to mix, even in the parking lot.

Those faithful to the Ford brand have a similar animosity toward the competition. Knowing that the price of a newer used truck would cost more than repairing Good Ole Reliable, I lobbied hard for Billy's right to retain the Chevy, but the girls began an all-consuming search for the perfect truck. Billy would soon learn that they settled on a Ford F150 extended cab.

"I swore off Fords in 1953," Billy said as we talked about his daughters' search. "That old 46 Coupe I had wadn't nothing but a lemon. And ya know Ford trucks are just terrible on tires. They will just wear a tire out in no time."

I nodded. I've heard this statement before.

At the time, Billy didn't know that his next truck, one of only three vehicles he has owned during his lifetime, would be a tank of an F150, an extended cab with dual-passenger doors that ate tires (and gas) for breakfast.

Billy said, "But you know, every now and then I've got to let my girls have to have their way. I reckon if they believe that the Ford will make me a good truck, I can give the company another try."

Good Old Reliable stayed in the shed, just in case. She isn't for sale. Don't bother asking.

To qualify as a work truck, Billy wanted to transfer and realign the wooden side rails from Good Ole Reliable to the fancy bed liner-equipped new Ford. He may have parted with the Chevy, but he wanted the railing to stay. Side rails are a necessity for farm trucks. Rails allow workers like Billy to secure their load and stack hay and other items higher than the standard height of the bed. I have never asked, but judging by the rust on the corner hinges, I guessed the side rails were at least thirty years old.

Janet drove from Charlotte, North Carolina, determined to equip the F150 with the proper appendages. Forty-eight hours later, she and Billy had finished the project. The Ford boasted freshly installed side rails painted jet black.

As the daughter of a Chevy loyalist, I pessimistically surveyed Billy's new mode of transportation. Walking around the carport, I mentally noted several less-than-desirable qualities: it barely cleared the carport ceiling, an extended cab and longer bed made parking a challenge, the interior technological gadgets were too highfalutin', and a cloth interior wasn't exactly what one wants to crawl into while wearing manure-laden overalls.

Pasting a congratulatory smile on my face, I said, "That sure is a fancy truck."

"Here," Billy said quickly, while placing the keys in my hand. "You drive."

Shaking my head while returning the key, I declined.

"Oh, no," he insisted. "This is your truck too. You go ahead."

Climbing into the Ford, I adjusted the mirrors and tried to back the F150 out of the carport. Immediately, there was a problem. The wooden rails that Billy could not live without blocked my view. Despite tilting the rearview mirror in a variety of angles, the only thing it reflected was the image of a newly painted sideboard. Exiting the truck, I readjusted the electric seat to accommodate Billy's longer legs and said, "Sorry, I can't see out of the back."

ꝯ

Janet lives hours away. For prodigal children who return for visits, there are not enough hours in the day to spend with family members, reconnect with lifelong friends, and, in Janet's case, complete another project on the still-unnamed Ford. Bless Janet's heart, it seemed like every time she visited for the weekend, she spent a majority of her time tinkering with that truck.

Now there's an idea: let's call the truck "Tinker."

I found Tinker undergoing another surgical procedure. A red plastic milk crate served as a stepstool. Orange drop cords snaked up Tinker's side and into the bed. A metal bucket full of nails and screws threatened to topple from the tailgate. Two crowbars and the hammer were lying in the bed. Frustrated voices escaped through the spaces in the rails.

"Here's Zippy," Janet said the moment she saw me. "She'll agree with me."

Outnumbered by women, Billy should have immediately thrown up his hands, but he did not.

Walking behind the truck, I encountered what my mother calls a "gualm." This made-up word defines the biggest, messiest, most pathetic attempt at a project in the history of all projects. Gualms require the use of every power tool one possesses, take ten times longer than planned, and end when someone resolves never to do that again. I have been making gualms for decades. From baking cookies to craft projects, I'm a messy gal when in project mode. This is why Billy and I get along so well.

"I tried driving this thing yesterday," Janet explained. "I can't see out of the back."

I responded with a slow, knowing nod. Translation: I am on your side.

Instead of removing the entire wooden plank that blocked Janet's view, Billy had taken the Skil saw and cut out a hole. Sawdust covered the bed and littered the lawn. Both workers were frustrated.

Yup. This was a grade-A certifiable gualm.

While placing her hand on the shiny black board she had just painted a month earlier, Janet said, "Zippy, this entire piece of wood has got to go…don't ya think?"

Since you asked, let's solve this problem with two little words: FOR SALE. I will get the cardboard, a Sharpie, and duct tape. The Ford can go. *The railing can stay.*

I didn't express those thoughts aloud. Instead I said, "Of course the board has got to go." Then I climbed into the back of the truck.

"See, Daddy," Janet said while putting her hands on her hips. "See. Zippy agrees with me."

While Janet and I searched for the drill with which to remove the screws (it had been conveniently misplaced), Billy argued that the middle plank was necessary. Janet and I countered that the bulk of the weight fell on the corners and that the middle board was not "load bearing."

It helps when women use technical terminology, just to prove that we girls do know a thing or two about trucks.

"Besides, what in the world would you stack in here that's higher than the rearview mirror?" I asked.

"That's right, Daddy," Janet agreed while examining the bed. "You're just hauling goats. Even then, they're secured in a dog crate." Smacking the board for emphasis, she said, "You don't need this rail at all."

"And you couldn't stack wood high enough to block the window," I said. "The truck couldn't haul that much weight."

It also helps to know a thing or two about stacking and hauling timber.

With the Chevy, a good load stacks about level with the top of the truck bed. Applying the same concept to the longer, larger, and much deeper Ford truck bed concerned me. Primarily because Billy and I can barely handle what Good Old Reliable hauls. A load of gravel is not too heavy or too much to offload; it's the same with hay, manure, and wood. Given Billy's tendency of making every trip count, I envisioned backbreaking shipments in the near future. The kind of trip where the bumper drags the pavement and I truly believe that I might die before we get the Ford unloaded.

Realizing that Billy was uncomfortable driving the truck and wanted me to drive it, I slumped my shoulders and said, "If I can't see out of the back window, then I guess I can't drive the truck."

A smile slowly formed on Janet's face.

Problem solved.

Days later, Austin Carter, a regular on the farm, visited. When he pulled in behind the carport, he immediately noticed that the Chevy was missing and asked if Billy had sold her.

"Oh, no," Billy said. "She's in the shed resting." He smiled and added, "Driving the Ford is kinda hard on the wallet. If you know what I mean."

Nodding toward his Cadillac Escalade pickup, Austin said, "Yup. I know what you mean. My truck's got a drinking problem, too."

The men circled Tinker. Austin checked the tires and then wrapped his hands around the side rails and gave them a hard pull. "Looks like you've got this one all set up for work."

"Oh, yeah," Billy said, walking around to the back of Tinker. "I'm fixin' to break it in."

While the menfolk perched a leg on the Ford's bumper and carried on with their manly conversation, I meandered through the carport, heading toward the garden, until a flash of silver caught my attention. Upon further inspection, I saw that Billy hadn't removed the piece of wood as Janet and I had agreed he should do. Instead, he had cut another hole in the rail large enough for me to see through and then installed a mangled piece of chicken wire over the back window. A mismatch of assorted screws and bent nails held the glass protector in place.

"What is this?" I demanded while pointing at the cobbled-up wire.

Smiling triumphantly, Billy said, "That's my head-knocker rack."

"A head-knocker rack? What in the world are you talking about?"

Billy explained, "Well Zippy, most folk call it a headache rack. You gotta have something that shields the window and the truck cab. So if the load I'm carrying shifts, what I'm hauling won't break the winder and knock me in the head."

Austin nodded. I should mention, however, that his Cadillac Escalade did not feature a head-knocker rack.

"I robbed a piece of wire from the chicken house and fastened it to the wood," Billy said.

"Has Janet seen this?" I asked.

Billy didn't answer. He and Austin exchanged a knowing smile.

Problem solved.

Edenwilde Book Club Soup

This recipe comes from Jo Anne Kilroy. She served it on the night Billy and I joined the Edenwilde Book Club for a lovely evening filled with good friends and tasty food.

Ingredients

1 quart chicken broth
1 (8 oz) container portabella mushrooms, sliced
½ cup celery, finely chopped
¼ cup onion, finely chopped
1 cup sliced carrots
2 teaspoons McCormick chicken seasoning
1 can mushroom soup
1 cup water
¼ cup dry sherry (or dry white wine)
¼ cup low-sodium soy sauce
2 cups cooked wild rice
1 tablespoon fresh parsley
1 tablespoon chopped oregano
1 cup cooked chicken, shredded

In a large container, add the chicken broth, onions, celery, chicken seasoning, cream of mushroom soup, water, sherry, and soy sauce. Stir well.

Add uncooked carrots to the mixture. Bring mixture to a boil, then reduce heat. Add rice and cooked chicken. Simmer for 25 to 30 minutes. Add fresh herbs just before serving. You can use dried herbs instead, but decrease the amount used since dried spices are more concentrated.

Simple Potato Soup

During the winter months, Billy and I eat a lot of soup for lunch. This recipe is quick and easy.

Ingredients
3 or 4 medium-sized potatoes, peeled and cut into cubes
2 quarts water
1 teaspoon salt
½ teaspoon garlic powder
½ teaspoon ground pepper
1 tablespoon butter
¼ cup shredded cheddar cheese (optional)

Peel and cube potatoes, then place in saucepan with water. Bring to a boil, and add salt and butter. Stir well. Reduce heat and simmer for 15 minutes. Taste potatoes to see if they have reached desired consistency. Add garlic powder, pepper, and more salt if necessary. Stir well.

Ladle into bowl and top with a pinch of cheddar cheese.

12

A New Tractor to Accessorize the Truck

One day I arrived at the farm just as Billy was leaving. "I've got to go pick up my new tractor," he said with a wave.

It is beginning to look like a contraption parking lot over at his place. A body can only drive one tractor at a time.

If owning a new truck wasn't enough, Billy committed to buy another piece of equipment he called a "TOB," which stands for This Old Bowen. Purchasing TOB increased the population to three: a Farmall, a Deere, and TOB. Billy also has, at last count, three hand-held tillers, two riding lawnmowers, a push mower, and an assortment of hand tools scattered throughout the property. Typically, all groundbreaking projects begin with the rounding up of necessary implements and, subsequently, fiddling with the choke, starter, timing belt, spark plug, clutch, or battery. Most often, these contraptions only crank after Billy sprays one of the parts with a liberal application of WD-40. You must also utter the ritualistic "hocus pocus," cross your fingers, hold your breath, and say a prayer prior to pulling a cord or turning the key on any piece of equipment.

No turnkey quick starts around here, not by a long shot.

If you grew up around farms, you recognize this routine. In other words, Billy's place is probably just like yours, only with a lot more critters and wayward tools.

There are times when I can reason like Farmer Billy. Each phase of the garden requires a specialized piece of equipment. He uses the sodbuster in the spring to break up the "faller" ground. Then a harrow, which he calls a "har," digs deep below the surface, turning and incorporating organic matter into the soil. A bull tongue works with the layoff plow to create straight rows that are ready for plants. Billy also fills the planter with seed and fertilizer. Driving this device down the rows is a lot faster than dropping seeds by hand. Most of the time, I drop the seeds into rows that are thick and imperfect.

"You get more plants in a crooked row than a straight one," Billy says, trying to make me feel better.

Tractors do not come fully assembled and dirt-ready. Each farmer must install a specific tool depending on the task. Billy has three tractors because each piece of equipment uses heavy parts and extra gadgets that require, at minimum, two strong men to remove and then refit the machine. For Billy, it's easier to crank a different tractor than change out the gadgets. I know because I have helped change out the tractor. It petrifies me to envision Billy working on his tractor by himself. They are dangerous pieces of equipment known for maiming and killing farmers at will. Billy's wonderful neighbor, Lamar, often helps. His friendship through the years has blessed us all. But Lamar can't be everywhere, and when Billy gets the notion to bust up some clay, he moseys out to the shed and starts changing out the tractor. He fishes beneath the seat for a wrench, ambles over to the plunder building for a can of WD-40, and starts beating the bolts loose. Dragging a har, a mule board, bull tongue, or some other odd-named implement, he tugs, lifts, heaves, and eventually connects everything he needs to break the new ground or knock back some weeds.

The first year I helped Billy, I stored my tools in a designated area inside the shed. While I can organize seeds according to alphabet, I am a cast-aside gardener when it comes to working in the field—one who lets the tools drop where they may and then wastes half the day searching for them. I am working to break this habit. My gloves are particularly shifty. I believe the fabric in the finger portion walks on its own accord. Camouflaged amid the grass and weeds, they snicker as I step over them, suppressing a guffaw when I ask, "Where did I leave my gloves?" Eventually I give up and work with my bare hands. Days later, I find the elusive protectors in nonsensical places, like draped across the chain-linked fence or hanging on two rusty nails.

I swear Billy has gremlins that come out at night and play with the tools. They also play with Billy's hoe. Almost identical in appearance to my Grandma Wonderful's, the once sharp edges of the hoe have worn smooth. The blade is worn down to a mere two inches across, making it the perfect gadget for hacking weeds.

"I'm just going to leave my good hoe over here," I announced as I hung the shiny implement from a ceiling beam. "Feel free to use it all you want, until you wear it down like yours."

Billy prefers mine anyway. My full-sized hoe lets him cover more territory in less time. But his nubby blade does slide perfectly into those hard-to-reach places like below the beans.

For those who have never used a hoe, let me explain how to remove weeds. You can either physically rip them from the earth, knock off as much of the dirt as possible, and then toss the uninvited guests in the sun to die (insert evil laughter while the weed shrivels and dies), or you can press the blade just beneath the soil and, while lifting upward, pull both the tool and the weed toward you (insert more evil laughter while the weed shrivels and dies). Usually, having a weed-free garden requires a combination of both processes.

The next few paragraphs are for first-time gardeners. A new gardener can invest a lot of money into growing her own food. Choose tools based on comfort, not popular trends and price. Petite gardeners like myself may find it necessary to remove a couple inches from the ends of their tool handles or purchase smaller, more efficient, pitchforks to work the soil. Tillers are wonderful, but I find them difficult to crank and control. Maintenance is expensive.

For many, raised beds might be the best way to start. Using wood, preferably cedar, construct a rectangular-shaped frame. Add a large amount of organic matter such as shredded newspaper and straw at the bottom of the frame. Moisten the paper and straw to expedite decomposition. Next, purchase potting soil. When it comes to this necessity, gardeners literally buy a pig in a poke; we can't see what we're getting, and these bags often contain little more than shredded trees. I have learned that the best way to purchase soil is to feel the bag. Pinch the corner. Do you feel large chunks and pieces of bark? If so, move along to a different brand. Good soil is soft and pliable, never chunky. The discount store, Fred's, sells an excellent, reasonably priced off-brand. After adding dirt to the bed, be careful not to walk in the area. Compressed dirt is difficult to work with.

Plants need food, either nutrient-rich compost, organic fertilizer such as manure, or 10-10-10. In areas with clay soil, a pinch or two of lime also works wonders. An application of one cup lime and one cup fertilizer should last several months. Just sprinkle the fertilizer between the plants, never touching the leaves or roots. Raised beds have the ability to produce high yields because the compact growing area allows gardeners to manage weeds and pests without backbreaking work—no tiller required.

Protecting your hands is a must, even if you enjoy the feel of the dirt on your hands. Each year I destroy at least two pairs of high-quality garden gloves. Finding adequate protection is increasingly difficult. Plain cotton gloves work well during the harvest season, but planting demands a sturdier, more rugged fabric that can withstand digging. I do not wear leather gloves. For me, they aren't practical. I need a washable glove. Ideally, the perfect glove retains its shape after washing. Alas, the fabric around the reinforced fingertips usually shrinks. Even though I have a petite body, I have large hands. Finding protection that extends past my wrist is difficult. When shopping, select a product that covers at least one inch above the bend of your wrist. Nothing is more annoying than dirt, or manure, finding its way inside your glove.

Some gardeners love the feel of the sun on their faces, but protective hats and sunscreen are essential as well. Many ladies with long hair find straw hats uncomfortable, opting instead to tighten a cap or visor around their heads. Sunscreen is equally important. Water BABIES® SPF 50, applied on the face, neck, ears, and hands (even though they are covered), is part of my morning ritual. I also don sunglasses while working in the garden. Despite this armor, my skin develops sunspots. Dirt passes through the fibers of my gloves and embeds in the ridges of my fingertips. My hair bleaches to an unappealing carrot-top orange. In the summer, appearances aren't as important as delicious vegetables.

Mexican Cornbread

Mexican Cornbread is served as a meal in the Albertson home. This recipe is a modified version of Betty Jane Tilley's recipe found in the Garden of Recipes *cookbook printed by the Hoe'n in Euharlee Garden Club.*

Ingredients
1 cup sour cream
2 eggs, beaten
½ cup vegetable oil
1 teaspoon salt
1 (8 oz) can Mexican corn, drained
½ cup cheddar cheese
1 cup self-rising cornmeal

Note: When using Billy's cornmeal, or other stone-ground cornmeal use 1 cup of meal then add ¼ cup all-purpose flour, 1 teaspoon baking powder, 2 teaspoons salt, and ½ teaspoon baking soda. Otherwise, the bread will not rise.

Preheat oven to 400 degrees.

Lightly grease a cast-iron skillet and place it in oven while mixing the ingredients.

Pour cornmeal into a bowl. Make a hole in the center, and add eggs, oil, salt, and sour cream to the hole. Stir well. Add the can of Mexican corn and the cheddar cheese. Stir well.

Remove skillet from oven and lightly sprinkle with cornmeal to prevent sticking. Pour mixture into pan and bake for 30 minutes or until golden brown.

Rice Pudding

Billy Albertson loves rice so much that he purchased a rice cooker. While he toils in the garden, the cooker works its magic in the kitchen.

Ingredients
½ cup uncooked rice
1 tablespoon cornstarch
2 eggs, separated
2 cups milk
½ cup raisins
½ cup sugar
Pinch of salt

Cook rice according to package instructions.

After rice is thoroughly cooked, transfer it to a saucepan. Mix eggs, milk, and cornstarch together and pour into pan. Stir well, and cook for 2 minutes. Add sugar, raisins, and pinch of salt. Serve warm.

Spinach Dip

This little dish is a favorite at our house. I make it every Saturday during college football season. Substitution note: Three cups of fresh kale or spinach is an excellent substitute for frozen.

Ingredients
10 ounces frozen spinach, thawed (keep juice)
½ onion chopped
4 ounces cream cheese
¼ cup milk
1 tablespoon chopped garlic
1 cup mozzarella cheese
Salt and pepper to taste

Place cream cheese in baking dish. Add milk and garlic, and stir well. Mixture will be thick. Add spinach and onion. Sprinkle with salt and pepper. Top with cheese. Bake at 350° until bubbly. Serve with corn chips.

13

Pesky Garden Pests

Most visitors to Billy's farm have the ulterior motive of obtaining some of his delicious homegrown vegetables. Visitors of the pesky variety—raccoons, crows, geese, hawks, insects, rats, and squirrels—have a similar mission. Staying on top of the constant influx of pests requires diligence. Left uncontrolled, these varmints wait for Billy to fall asleep, then tiptoe into the garden and eat everything he and I have worked hard to grow.

Those who believe that all animals are cuddly and can do no harm might want to skip this chapter.

I will begin with the crow. Crows are opportunist omnivores that will eat just about anything. They are cunning creatures whose sharp eyes know that springtime gardens yield tasty treats. For them, corn is a veritable all-you-can-eat buffet. After green leaves appear, crows walk the rows, grabbing the corn and pulling up unfurling shoots. Calling to the rest of their flock, they gobble a savory kernel and then proceed to the next shoot and the next until their appetites are satisfied.

For the record, it takes a lot of corn to satisfy a crow.

Replanting is a waste of time. Once a crow finds your garden, he tells the entire rookery. A rookery is a community of crows tasked with the duty of protecting their own and expanding the crow kingdom. A murder is a large flock of crows whose plan is world domination. Corn does not satisfy these evil ones; they also devour innocent hatchlings. They are watchful beasts. From lofty vantage points, crows wait, listening to the innocent, precious peeps of songbirds. They wait patiently while exhausted parents such as cardinals and sparrows search for food, and then the murderous thieves snatch defenseless, featherless baby birds. Carrying their prize, they rip the baby birds we adore into tiny bite-sized pieces and feed them to their own fledglings, murders in the making.

Don't talk to me about survival of the fittest. I have no sympathy for a crow. None.

Several natural methods exist to help farmers control crows. These include providing decoy food (feeding them all they can eat in an area that is far away from your garden); hanging pie pans and compact discs in the area you want them to avoid; spraying them with a water hose; animated crow killing calls; and clapper devices that emit a noisy sound that is said to scare away birds. Then there is the scarecrow, which, while cutesy and fun, is about as effective as the previously mentioned methods.

On the other hand, Avitrol® is a chemically treated grain that, when consumed, makes crows behave a little coo-coo. Crows who consume this chemical give warning cries that frighten other birds. Usually a crow that consumes Avitrol® dies.

In my opinion, a BB gun or rifle filled with birdshot provides a more humane method. Aim. Shoot. Hang 'em high as an example for others.

According to a segment on PBS's *NATURE*, scientists believe that crows have the ability to remember which human is friend or foe. They also communicate with each other. Maybe you have noticed a scarecrow holding what looked like a dead crow. These are common in my hometown. Perhaps you've wondered about the significance of hanging a carcass in the garden. Since this bird is particularly intelligent, it knows that scarecrows are not human. The crow is not afraid. Amused, yes. Scared, no. However, hang one example from the inanimate, outstretched arms of a scarecrow, and the rest of the flock quickly gets the message. Word travels through the flock: mess with the corn planted here, or the baby birds, and you're a goner. This farmer means business.

Knowing this, it only made sense for me to slam on the brakes, put my husband's vehicle in reverse, and pull over on the side of the road when I saw a dead crow lying on the sidewalk.

I was doing the city a favor, my civic duty no less.

The bird was doornail dead, lying outside a guarded, gated community where a certain reality show is filmed. You can bet your bottom dollar that nary a woman inside the mansion would come near a crow, live or dead, with a ten-foot pole.

I found a plastic bag in the trunk, rushed to the sidewalk, scooped up the crow, and drove to Billy's.

Exiting the car, I noticed three crows strutting their stuff down the corn rows, bending, gobbling, noshing away. I had arrived just in time.

"Where's your BB gun?" I asked Billy.

With a nod toward the shed, he said, "Propped against the tractor."

"Got my eye on three crows, but they'll fly before I get there." I stuck out my hand and shook the plastic bag. "Brought you a dead crow. Picked him up on Bowen Road."

You would have thought I gave the man a sack of hundred-dollar bills.

"A dead crow." He slipped his feet into a pair of shoes parked at the door. "Well now, I'll put this old boy right to work. He'll scare off his brothers quicker than you and the BB gun."

After you've dealt with crows, rats begin their assault. Animal food attracts rodents. Since Billy purchases large quantities, protecting animal food from an infestation takes work. One female rat can birth up to ten litters a year. With each litter averaging six to eight babies and those babies having babies, controlling rodents is a priority on any farm. Billy stores animal food in metal trashcans with strong, weighted-down lids. Plastic containers are ineffective. Rats use their sense of smell to find food. Once they locate a meal contained in plastic, they simply locate a weak section, usually the lid, and gnaw their way inside.

Believe it or not, Billy Albertson doesn't own a cat. He is in the market for a working cat, not a domesticated Garfield. Until then, several traps work overtime to keep the plunder building almost rodent free.

Geese are also big pests on Billy's farm. Protected by the Migratory Bird Act of 1918, these migratory birds are troublesome because they no longer return to Canada each year. Migration is exhausting, you see, what with the wing flapping and the dangers of jumbo jets. The protected problems are now resident Canadian geese, permanent fixtures that are fat, aggressive, and reproducing rapidly. Each goose deposits two pounds of potentially parasitic droppings every single day. This poo can contain fecal coliform. Geese droppings add inorganic nitrogen and phosphorous to freshwater lakes. Their waste stimulates algae growth and depletes oxygen in the water, which harms aquatic life.

The only way Billy and I can keep them out of the garden is by running through the rows while screaming like banshees. Well, I run through the garden screaming like a banshee. Billy laughs. Occasionally he lights firecrackers. That usually sends the geese squawking over to the neighbor, who lights his own noisemakers and sends the flock back to Hardscrabble Road.

Every poultry owner must deal with wicked raccoons, elusive foxes, and villainous hawks. All of these beasts stalk defenseless chickens. They also stalk human behavior patterns. They know when you feed and when you sleep. They strike when least expected, even in the middle of the day. One particular year, Billy watched helplessly as his flock of chickens dwindled from thirty-five to twenty, then to seven. The carnage happened fast, in two days. Scouring the woods, I found the indisputable proof of paw prints. Billy was downtrodden. Facebook friends who had never met Billy rallied to help him. Bob installed cameras and Michael set traps while Billy and I triple-checked the coop, reinforcing the wire.

The motion-sensing camera captured images of a gathering of rats that were dining on chicken feed under the cover of darkness. But Michael's traps snared a different culprit the first night, a raccoon.

"Set the trap again tonight," I told Billy. "Where you have one, you have ten."

Two miles down from Billy's, a developer had drained the lake and then used a bulldozer to remove almost all of the trees on the property. Apparently, the area needs another bank and several offices more than the wildlife need a home. This type of urban sprawl has diminished the habitat of foxes but caused an overpopulation of raccoons and hawks.

Raccoons are scavengers. Thriving on opportunity and relying on human laziness, they have the hobby of dumpster diving. Mixed-use developments are convenient to humans and heaven for coons. Discarded, half-eaten food provides a smorgasbord of nibbles. After noshing on partially eaten burgers and fries, it is nappy time. Where better to catch a few winks than in the attic of a house in the nearest subdivision? After ripping insulation from the walls and rearranging it into a comfortable bed, a female raccoon invites her boyfriend over for some snuggle time.

We all know what happens next: babies in the belfry.

While bear-proof receptacles are common in national parks and other areas near active bear dens, municipalities have overlooked the damage a single raccoon can cause. These strong creatures have the ability to penetrate metal cages, including the eave vents of a house. Nothing short of an electrical fence will keep them away from chickens or your attic. Cities should control the population of this masked bandit by requiring, at minimum, rodent-proof trash receptacles.

Farmers also know that once a fox gains entrance into the henhouse, it returns each night at approximately the same time and removes a single chicken. There are laws against shooting foxes. Posting an ornery mutt is perhaps the best deterrent. Hawks also stalk the innocent and defenseless. Swooping in without a sound, they fly around the coop, terrorizing panicked poultry. Upon recognizing the Grim Reaper, chickens attempt to escape their cage. They run the perimeter of the coop, pressing their heads through holes in the wire, and then BAM! The hawk claims its first victim.

When my Grandpa Winchester's health began to decline, home health nurses visited. On a particularly sunny day, he and the nurse were on the porch. She was taking his blood pressure when a hawk lit on a power line.

"I'm gonna kill that thing," he said, using a matter-of-fact tone that gave both my grandmother and the nurse a chuckle. "Soon as you finish checking my blood pressure I'm gonna blow his feathers clean off."

"Ya think so?" the nurse responded.

"He can't hit that hawk," Grandma Wonderful said with a snort.

Grandpa excused himself and returned with a Winchester rifle. Taking a seat in the rocking chair, he rested the rifle on the banister.

Grandma Wonderful and the nurse smiled. There was no way he could hit the hawk. It was too far away, across two lanes of traffic, for Pete's sake. Grandpa was too old, too shaky. Even the hawk was confident. Sitting on the line with its feathers all puffed out, taunting. I am sure he never even felt the shot.

Before you rosin the violin's bow, note that these predators are cold-blooded killing machines. Grandpa wasn't running a food bank for wayward hawks. He was protecting the chickens and his livelihood. There are plenty of squirrels and chipmunks for hawks.

After Billy's colossal chicken catastrophe, he determined that he needed to improve his marksmanship. I drew a target on a paper plate, and we both commenced to practicing. A word about gun safety: we use a BB gun for target practice and critter eradication. We're not blasting up the residential area; we're merely controlling the critters that dare press a toe across the property line. I believe in a homeowner's right to protect his poultry from the likes of anything that might eat it, particularly something that could carry rabies. Billy squinched one eye closed and squeezed the trigger. The paper popped. A hole appeared just below the circle.

It should be noted that, in a previous life, I lived in Sevierville, Tennessee, the heart of the Smoky Mountains, in the backyard of where my people once roamed. I was enrolled with a group of future leaders called Leadership Sevier. Our class visited some of the best companies in the area, where we participated in lectures, classes, and team-building exercises. During a trip to the Great Smoky Mountains National Park, the group learned about trail maintenance and about invasive species such as the woolly adelgid that destroyed a large majority of the hemlock trees. We also learned how folk used to hunt bear long ago. Using a blowgun that was approximately three feet long, the speaker demonstrated the technique and then asked for volunteers. Even though I didn't raise my hand, he pulled me forward.

I've been pulled, or pushed, to the front for most of my life. A girl gets that a lot when she's short.

"You've got to take a deep breath and put a lot of air behind the dart," he told me. "Then, bring the blowgun to your mouth. You need a tight seal and a lot of air, or the dart won't hit the mark. But don't put your mouth on the gun while you inhale, or you'll suck the dart down your throat."

Nodding, I stood behind my male counterpart who had actually volunteered, the one with years of hunting experience and a deer head on his wall at home. He stood where instructed, did as told, and hit the fake bear right in the shoulder.

Anything he can do, I can do better.

It's pretty much impossible to hold something that's almost as long as you are tall. The reed felt heavy, difficult to manage. I brought it to my face and winked one eye closed, then switched eyes for a better look. I inhaled and blew out hard. A collective groan escaped from the men. They cupped their hands protectively over their crotches, and a couple turned away.

"Lord a mercy," someone said.

"Son of a...," said another.

I hit the fuzzy fake bear right in the baby-making department. It wasn't a clean kill, not by a long shot, but I'd lay dollars to donuts that the bear wouldn't bother me again.

Needless to say, the menfolk treated me differently after that team-building exercise. My point to this sidebar is that, odds are, whatever was killing Billy's chickens was safe. Instead of trying to shoot the culprits, the best plan of attack was probably moving the chickens to my house until the coast was clear. And that's what we did.

It's possible that some of your people have enjoyed squirrel dumplings. We just don't talk about it at dinner parties. Not so long ago, hunting squirrels was a family tradition. Boys of all ages entered the woods on a rainy day, when moisture muffled their footsteps. Grandpa, father, son—three generations of hunters spent the day walking the woods. Their fingers itched to switch off the safety. They listened for that telltale bark that comes when a squirrel sounds the alarm.

Today, squirrels are taking over the world.

If you ask me, the creatures are nothing but beady-eyed rats with puffy tails. Squirrels love to raid birdfeeders. Sitting inside the feeder while using their tail as an umbrella, they scoop seeds into their evil little hands and nosh until they become morbidly obese. Then they waddle to your rooftop, hang upside down by their claws, and gnaw holes in the side of the house. After gaining access, they absolutely adore building nests in attic insulation. Pulling and tugging, they gnaw through wires, insulation, support frames, and anything else that gets in their way.

For those who believe squirrels are cute and cuddly, let me say that they are the most destructive creatures you can have near your home. Having a family of squirrels living in the attic is dangerous and unsanitary. They scamper and romp. They are awake at 2:00 A.M. gnawing and scratching fleas. Their escapades reverberate through the house with a rapid thump, thump, thump, scratch, scratch, scratch. Their entrance holes provide a welcome mat for other creatures such as raccoons and woodpeckers.

Before the hate mail arrives, I should share that I grew up nursing baby squirrels with bottles and dangling worms from tweezers so the baby birds wouldn't die. Working for the power company gave my dad access to orphaned animals whose parents had touched the wrong wire. My parents taught me about nature, particularly about balance. Right now, we are out of balance. Trust me. When a raccoon or, heaven forbid, a squirrel starts gnawing on your house, you will either level a bead and pull the trigger, or call a pest control company. At some point, controlling the pest population becomes a necessity. Either they go, or your peace of mind goes.

Billy doesn't need an exterminator; he has a neighbor who despises squirrels, and that neighbor is a darned good shot.

ଋ

The metal screen doorframe rattled against the Plexiglas, announcing my arrival. When summer has faded, I try to visit Billy at least weekly, just to catch up on how things are going and warm myself near his woodstove. On this particular day, a heady aroma escaped the kitchen and filled the tiny den.

"I want you to have a bite of this," Billy said as he extended a steamy bowl toward me.

Upon examination, I quickly declined and said, "You're not fooling me. I grew up in the mountains; I know squirrel dumplings when I see them."

Billy laughed and shoveled a bite into his mouth. "Nothing wrong with a little squirrel for dinner. You know they've gotta be healthy. They've been eating my corn all summer."

Chicken and Dumplings

This recipe is from the cookbook, A Taste of Pilgrimage, *printed by the Pilgrimage United Church of Christ. When Billy was growing up, chickens were too valuable to eat. His mother sold eggs and young hens to city folk so she would have spending money. Billy grew up eating squirrel dumplings, which you, if so inclined, can substitute in this recipe. The church's book club, incidentally, is one of many that befriended Billy. The recipe comes from Phyllis Kincaid.*

Ingredients
2 bone-in chicken breasts
2 quarts chicken broth (reserve 1 cup for batter)
2 cups flour
1 teaspoon baking powder
1 teaspoon salt
1 egg

Cook chicken breasts in 2 quarts of broth.

When cool enough to handle, remove chicken from bone and shred into bite-sized pieces. Reserve broth. Mix flour, baking powder, salt, and egg. Ladle one cup of hot broth into dumpling batter. Mix well. Roll out and cut into squares, or drop by teaspoonful into boiling broth.

Cook 5 to 8 minutes.

Add chicken to pot and serve.

Cheese Straws

Cheese straws are a staple at Southern gatherings. This recipe comes from the Garden of Recipes *cookbook printed by the Hoe'n in Euharlee Garden Club. This lovely group of ladies has won multiple awards and works hard to keep their community beautiful. I enjoyed their cheese straws when I visted their group in the charming Cartersville area. This recipe is from Janet Martin.*

Ingredients

1 cup all-purpose (plain) flour
1 ½ teaspoon baking powder
½ teaspoon salt
½ cup shredded cheddar cheese
2 tablespoons plus 1½ teaspoon cold butter
⅓ cup milk
2 teaspoons paprika

Preheat oven to 375 degrees.

In a small bowl, combine flour, baking powder, and salt. Stir in cheese. Cut in butter with a fork or pastry knife until mixture resembles coarse crumbs. Gradually add milk, tossing with a fork until dough forms a ball.

On a lightly floured surface, roll dough into a square. Cut in half and cut each half into ½-inch strips.

Sprinkle with paprika.

Spray baking sheet with cooking spray. Add cheese straws and bake in preheated oven for 6-8 minutes. Serve warm.

Crock-Pot Macaroni and Cheese

This is another perfect dish for Billy, who only needs to throw the ingredients together, step outside and work, and then come home to a hearty pot of creamy mac and cheese.

Ingredients

8 ounces of elbow macaroni, cooked al dente and drained
1 can evaporated milk
1 cup whole milk
½ stick of butter (melted)
2 large eggs, beaten
4 cups grated sharp cheddar cheese
Salt and pepper to taste

Tip: Do not overcook macaroni. It will continue cooking in the Crock-Pot.

Pour the cooked noodles, milk, and melted butter into the Crock-Pot. Beat eggs and pour into pot. Add salt and pepper. Stir well.

Add three cups of cheese and stir. Then add the final cup of cheese.

Cover and cook for 3 hours on low.

14

Gaga *for* Goobers

Most women enjoy shopping and can't wait to hit the mall for the latest sale. Me, I prefer a good old-fashioned hardware store. Hardware stores smell like work, sweat, and pride. If you want to feel the pulse of a community, visit a locally-owned farm supply store. Friendly folk and a floor stocked with 100-pound seed bags feel like heaven.

Each spring I visit Ladd's Farm Supply in Cartersville, Georgia. A trip to this family-owned business is worth the two-hour drive from Atlanta. One-third of the store displays various types of seeds available for purchase. Metal and wooden bins house loose seeds, measuring scoops, and paper bags. The seed area is a help-yourself type of place. Of course, they'll measure the seeds if you wish, but employees at Ladd's also recognize customers who must run their hands through the bins, caressing the seeds. Customer service is important to the staff at Ladd's. Someone will answer your questions and load your vehicle. They will even order anything you might need for the farm or garden. They mirror the honest, hard-working values of long ago. Perhaps that is why they trust you to scoop your own seeds.

I marveled at this opportunity, knowing that some hardware stores designate a staffer who spends all day behind the counter. Almost giddy, I buried both hands in the bin, shoving them in deep, way past my wrists. As I touched as many seeds as possible, my intent wasn't to embarrass my daughter…but I did.

"Mom. What are you doing? Can we just get what we came for and go?"

I needed a moment. Seeds comfort me. With their promise of treasures to come, touching their hard, smooth shells filled me with hope.

A door chime announced the arrival of another customer. Staff greeted them with a smile, called them by name, and asked, "What can I do for you today?"

To expedite my purchase, Jamie opened a small paper bag and placed a scoop in my hand. Scanning the multiple compartments, I was overstimulated with possibility.

"I should get some half runners," I said to Jamie, who took the scoop from my hand and filled the bag. After sealing it, she located a pen conveniently hung above the bins, popped the cap, and tried to write on the bumpy bag.

"That will work better if you write on the envelope before we fill it."

Jamie rolled her eyes, emptied the bag, and began again.

"Oh, look," I said while moving to the next section, "they have pink half runners. Look how pretty the seeds are. And rattlesnake beans. I don't even know what rattlesnake beans taste like, but we must get some."

I was rambling, drunk with bean love.

Skipping past the blue lakes, I said, "I don't really like these. They grow too close to the ground. Even I have to stand on my head to pick them."

Paper envelopes bulged. Glue threatened to release the bottom seal as I moved through the sections, scooping, smiling, and taking photos with my camera. I never leave home without my camera. Life deserves recording on film.

"Mom. Do you think we have enough?" Jamie asked, impatient.

Turning to see her arms full, I said, "I guess we could always come back."

As soon as we stacked our haul on the counter, the checkout process came to a screeching halt when I noticed an open bag with the words SEED PEANUTS, GOOBER PEAS. I turned to Jamie and said, "Lawd have mercy, they have peanuts. Grab another bag."

Farmers in the state of Georgia proudly produce 49 percent of all the peanuts grown in the United States. According to the Georgia Peanut Commission, the 2009 harvest weighed in at 1,782,650,000 pounds of goobers. That's just over a billion pounds, folks.

With numbers like that, I figured Billy and I could at least coerce a couple of pounds from his soil. Barely containing my excitement, I felt my heart pounding and my smile widening as peanuts tumbled into the bag. I could not wait to get these goober peas in the ground.

As Mr. Floyd rang up the sale, I announced, "Today you have made me the happiest gal in the entire world."

He nodded and said, "Thank you, ma'am. We try hard."

I really wanted to leap across the counter and hug his neck, but I had embarrassed my daughter enough for one day. I could return later for hugs.

Outside, I dialed Billy's number and said, "Crank up the tiller. I've found us some peanut seeds."

ꝋ

It isn't necessary to travel two hours for seeds. Raw peanuts purchased at the grocery store will suffice. Cultivating peanuts is easy. Because of the long growing season, peanuts are perfect for container gardens. Commonly grown in the flatlands of Georgia, the plants acclimate to a harsh growing environment. The foliage is lovely, reminding me of clover, a shamrock, and the leaves of a mimosa tree all rolled into one. Each plant grows up to eighteen inches tall and then surprises us with tiny bright yellow blooms. The edible part grows below ground. A month after flowering, these self-pollinators began making babies.

To plant goober peas, dig a row and place kernels two inches apart and two inches deep. Shelling isn't necessary when using store-bought peanuts. Bury the shell and all. Seedlings emerge in ten to fourteen days. Peanuts need an inch of water each week in order to produce fat kernels. Approximately five months later, it's harvest time.

Commercial harvest is a two-part process. Farmers attach a digger to their tractor and maneuver it down the rows. This tool lifts the plant from the ground, shakes it, and then turns it upside down, leaving the plant to dry for a couple of days. Later, a combine collects the dried peanuts, detaches them from the vine, and drops the vine back to the soil. The goobers are then cleaned, shelled, dried, and transported to manufactures that produce yummy, high-protein foods. Peanuts that travel directly to farmer's markets, shell intact, are called "green" or "raw." The terms are interchangeable.

ꝋ

Walking our peanut rows, Billy nodded and said, "These goobers are doing a great job."

I fell in beside him. Below-ground crops require patience. I wanted to sneak a peek, to scratch the earth just a tiny bit and view the progress. The best I could do was cross my fingers and hope for the best.

Since Billy and I planted only a handful of seeds, we waited until after the vines faded and turned yellow. When our harvest day arrived, Billy pressed a pitchfork into the soil, lifted the vine, and noticed something horribly wrong.

Rats.

Rats had eaten most of our goobers, leaving us a scant handful to enjoy. This particular vector loves underground crops. Rats scratch the surface, gobble down goobers, and move to the next plant. Because the primary root system remains undisturbed, the plant continues growing—never even wilts—leaving farmers unaware of a limited harvest until it's too late.

"Well, well," Billy said as the pitchfork yielded empty vines. "I'll be dogged."

This is just one reason why a farmer is the biggest gambler around and why pesky pests must go. The rats didn't care that I loved these peanuts or that I had worked hard and weeded the patch in the hot sun.

Billy needs a ball-bearing rat trap of the feline variety, an alley cat who will kill these critters and proudly carry them to the doorstep.

Now that our harvest boasted only a few goobers, we determined to eat them raw.

"My sister Betty loves boiled peanuts," Billy said. "I had hoped to boil her up a batch, but I reckon we won't be making any this time around."

Green peanuts have a distinctly different flavor from dry roasted or boiled. Sweet and moist, they are delicious when consumed shortly after harvest.

"I remember the first time I tried peanut butter," Billy said. "I was in the third grade at Birmingham Elementary School. Miss Newton announced that she had a big surprise for us. She told all of us to bring a slice of bread with us to school the next day, and we'd get to try something called peanut butter. Well, our folk didn't have sliced bread, so I brought a biscuit. I can still remember her smearing peanut butter on my biscuit. Let me tell you, that was the best thing I had ever put in my mouth."

Memories like those are why we grow peanuts. Farming and cooking bind us together like peanut butter stuck to the roof of your mouth.

Twice Boiled Spicy Peanuts

Popping up along rural roadsides across the South, boiled peanut stands provide tailgaters with the perfect football snack. Boiled peanuts, whose pronunciation varies from "balled" to "burl'd," depending on geographic location, are a traditional part of pre-game tailgating menus. Whether served warm or in a brown paper sack, peanuts are a ritualistic snack for true sports.
Note: For very spicy peanuts, double the amount of peppers and add the seeds.

Ingredients
3 pounds green peanuts
6 quarts water
1 lemon, sliced in half
1 garlic clove, cut in half
2 jalapeno peppers, cut lengthwise in half
2 bay leaves
1 tablespoon vinegar
½ cup salt
½ package crawfish boil (reserve remaining package)

Day One: Rinse peanuts well before boiling to remove any traces of dirt that might remain on the shell. Place in a large pot. Add water and all of the spices. There is no need to slice peppers, garlic, and lemon into tiny pieces. Quartering them is sufficient.

Bring the mixture to a boil. Cover and boil for 8 to 10 hours. Allow to cool on stove, then place in the refrigerator overnight. (You can serve them immediately, but when allowed to sit in the brine overnight, peanuts absorb more flavor.)

Day Two: Taste one peanut to determine if the spices are adequate. It may be necessary to add more salt and the remaining crawfish boil seasoning bag. Bring peanuts to a boil again, and simmer until you are ready to serve. Add extra water as necessary to prevent burning.

Discard brine before serving.

Peanut Brittle

This recipe takes approximately an hour and a half to prepare, cool, and break into pieces. It is delicious and well worth the effort.

Ingredients
1 ½ cups raw peanuts (shelled)
2 cups cane sugar
½ cup cold water
1 cup Karo light corn syrup
½ stick butter for recipe
¼ stick butter for greasing the baking sheet
½ teaspoon baking soda
½ teaspoon salt
½ teaspoon vanilla
Candy thermometer
Greased spatula
Heated baking sheet

Note: Pre-measure all ingredients and have them ready beforehand. Grease a baking sheet with ¼ stick of butter. Make sure the entire area is covered. Place cookie sheet in the oven on low heat to keep area warm. This allows for easier spreading.

Attach a candy thermometer to a boiler pot. Then add sugar, water, and corn syrup. Heat on medium-high until mixture begins to boil. This takes about 10 minutes. Stir just enough to keep the mixture from sticking to the bottom. As the mixture reaches 250 degrees, known as the hard-boil stage, stir in the butter and the peanuts. Continue stirring until mixture reaches 300-310 degrees. Be careful not to scorch peanuts. Mixture will turn a caramel color. Remove from stove and stir in baking soda. Mixture will foam up. Immediately pour the candy onto a warm baking sheet and spread a layer thin with an oiled spatula.

After candy cools completely, break into pieces and store in an airtight container.

Peanut Butter Balls

Grandma Wonderful made these scrumptious concoctions each year for Christmas along with rum balls (recipe follows). She arranged them on a tiered serving dish that was elegant and dainty. Traditional recipes such as these replay wonderful memories with each bite. I'd love to know about your family traditions.

Ingredients

3 pounds powdered sugar
2 pounds peanut butter (can use crunchy if you'd like)
1 pound butter, softened
Dash of vanilla
2 (12 oz) packages chocolate chips
½ bar paraffin wax
Several sheets of wax paper

Mix together peanut butter, butter, and vanilla. Slowly add powdered sugar. Mixture will become stiff. Form into round balls and chill for several hours or overnight.

Using a double boiler, melt chocolate chips and paraffin wax. Dip the chilled peanut butter balls into the chocolate. Place on wax paper. Allow chocolate to harden before serving. This recipe freezes well.

Rum Balls

Rum balls had their own place of honor on the serving dish. Raised a devout Baptist, Grandma Wonderful usually sent one of her children to the liquor store to purchase the main ingredient in this recipe.

My cousin, Jerry Cline, works at the local ABC store in my hometown. He deserves an award for outstanding customer service. When the holidays roll around, he knows the folk who can't be caught dead in the liquor store. He will even have your ingredient ready if you call ahead. I love to watch folk arrive during their annual trip to the ABC store. Jerry calls each customer by name and usually says, "Your momma uses the rum on aisle three." Perhaps he should share a recipe or two. I bet he has some good ones.

Ingredients
2 ¼ cups vanilla wafer crumbs
1 cup chopped nuts
2 teaspoons salt
½ cup bourbon or rum
1 cup powdered sugar
3 tablespoons cocoa
2 tablespoons Karo corn syrup

Combine vanilla wafer crumbs, salt, cocoa, and nuts. Mix in a food processor, and then place in large bowl. Pour ½ cup rum in bowl. Add corn syrup. Mix well.

Shape into balls, then roll in confectioner's sugar.

Refrigerate and allow flavors to blend. Best when served several days after making.

raisin'
Cane

15

You Gotta Have Friends

Not so long ago, there was a time when having friends was important. It was a time when America's way of life was threatened. Food was scarce. Provisions were hard to come by. Every available resource—gas, rubber, metal, fabric—went to the war effort. Turning to Americans, the government asked everyday folk and urban city dwellers to make sacrifices and do remarkable things during the worst of times. The government issued a call to action, and Americans responded.

"The Great Depression," Billy says, "was a *real* depression. We had our cattle and chickens. The depression was one of the few times Poppa and Momma had it better than most. Nobody had nothing, especially the city folk. They didn't have a way of making food, and people were desperate."

From this desperation, the Victory Garden emerged. Book IV of *The Victory Gardens Campaign* prepared by the US Department of Agriculture strongly encouraged community gardens.

> By growing and preserving your own supply of vegetables you reduce the demand on commercial stocks needed for our armed forces and, moreover, insure for your family an adequate supply of much needed protective foods.
>
> People living in cities and in metropolitan areas…should be encouraged to seek community plots or allotment gardens on vacant industrial property…and develop garden plots, 30 x 50 feet or larger. Town schools should develop school gardens planned and managed on such a scale that will provide large supplies of fresh and processed vegetables for school lunches.

Today, these victory gardens, also known as war gardens, might seem trivial. I promise you they were not. Gardens grew patriots. The

American way of life was under attack. The government challenged Americans to show pride in their country, and folk all over responded. Most families had someone serving in the military. The thought of soldiers going without provisions was unacceptable.

During this time, students across the country joined 4-H clubs and made this pledge:

> I pledge…
> My Head to clearer thinking;
> My Heart to greater loyalty;
> My Hands to larger service; and
> My Health to better living for
> My Club, my Community, and my Country.

An example of their hard work is evidenced in a report on 4-H members in Northampton County, North Carolina: "4-H'ers have pledged their hands…through the collection of 7,125 pounds of scrap iron, 2119 pounds of scrap rubber, 3,253 pounds of scrap paper, tin cans and records. Seventy-five percent of the members carried Victory Garden projects and 12,491 pounds of food have been canned by members the last three years of the war" ("Report on 4-H Members' Contribution to War Effort" [Northampton County], 1945, www.lib.ncsu.edu/resolver/1840.6/264).

Most everyone struggled during this period. Remarkably, they also pulled together. Finding fresh fruits and vegetables in the grocery store was impossible. The public food supply couldn't feed soldiers and keep up with consumer demand. People who had never touched the dirt purchased victory seeds. Then they rolled up their sleeves and made a difference. These Americans, now called "The Greatest Generation," lived out the saying that Billy often repeats: "To have a friend, one must be a friend."

They did more than plant a row of rutabagas; they fed their neighbors, strangers, and their families.

"We didn't go to the store for much," Billy said. "The only thing we needed was coffee, sugar, and flour. We'd get oil for our lamps and shoes once a year."

My, how times have changed. Americans love shopping, consuming, and then throwing away. We can no longer distinguish between want and need.

Food rationing was common for the Greatest Generation. "The city folk were really struggling," Billy told me. "We got the vouchers just like our neighbors and everyone else. Since we had such a big family we couldn't use them all. So we shared what we had with our neighbors. Back then, everyone worked together. Poppa knew that if we pulled together, everybody would be all right."

ꝏ

Zach Braswell is a stout young man who became Billy's friend moments after meeting him. Easily six feet tall with a sturdy frame and strong back, he is the perfect farm helper. After our initial introduction, I wondered, *Why would any high school senior voluntarily work on a farm when many other activities are available for teenagers?* Perhaps he was on probation and the work at Billy's was community service. While I admire Billy's trusting heart, life experiences have left me cautious and skeptical. *Yes,* I thought judgmentally, *I best keep one eye on the plow and the other on this new helper.*

Oh Zippy, you foolish gal.

After speaking to Zach, I learned that one of Billy Albertson's famous goat signs posted at the end of the driveway lured him into the farm. While walking home from school, Zach noticed the sign and stopped for a visit. That moment was a catalyst that morphed into volunteering. His intent was to help and perhaps spend a bit of time with the animals.

My lesson: judge not, and be more like Billy, who first sees the good in everyone.

"I tell ya, those signs bring in all kinds of customers," Billy said with a smile.

The signs also bring in a helper or two.

When I asked Zach why he kept returning to Billy's, a puzzled look crossed his face. "Hmm," he began, then paused. I secretly hoped he could help me understand the reason why we both work long, hot hours in the summer. "You know. I'm not real certain why I keep coming back. It's kind of a magical place."

After I knew Zach better, I learned that Billy is the closest thing he has to a grandpa. Billy fills that role for many.

Zach is an animal whisperer. Not only does he love animals; his home is also a Noah's Ark of small animal species ranging from ducks to pythons. He is a reliable pet sitter and has often bunny-sat for me. In perhaps the ultimate act of trust, Billy placed Zach in charge of several geriatric hens that now live at Zach's home. Additionally, Zach possesses an uncanny ability to know information typically reserved for old souls. An old soul is someone who, despite his or her chronological age, is emotionally solid, wise, and dependable.

Many adults worry about our youth. How will they afford college, a home, a family? Will they be able to feed themselves in this ever-changing world? Zach gives me hope.

Outwardly, Zach is a young man. Inside, he possesses a wealth of accurate information gleaned by doing instead of observing. I attached the title "Old Soul" to Zach the moment Billy said, "We need to sow us a bed of greens."

A partially harvested bed of greens needed replanting. After tilling the bare spaces, Billy instructed Zach to "go get a bucket of sand."

Questioning the instructions, Zach looked at me for guidance. Bewildered, I shrugged a response. I didn't know what sand had to do with collard greens either, but if Billy requested a bucket of sand, then a bucket he would get.

Billy took the container Zach brought and poured a small amount of sand into an aluminum pie pan that was so worn it threatened to break in half. Then he took a bag of collard seeds from me, thrust his hand inside, and retrieved the jet-black seeds.

"What you do here is pour the seeds into the sand," he explained with a sprinkle.

Billy's leathery fingers pinched. Kneading, working, blending seed with sand, he said, "Now we mix this up good."

Pausing long enough to point to the field, he said, "Go ahead there,

Zach, make a bunch of rows."

Using the hoe, Zach created what Billy calls "skip rows," which are areas planted to fill in bare places. Zach leaned against the hoe and asked, "Shouldn't we just sprinkle the seeds directly into the rows?"

I agreed. *What's this business about sand?* I wondered. *Why are we adding this extra step?* I also wondered how Zach, who has lived in a subdivision most of his life, knew anything about planting greens.

Billy shook his head. "No. The seeds are so small you need to mix them with the sand so you won't plant them too thick."

Zach and I exchanged another puzzled look.

Working side by side, Zach scattered sand and seed while I covered the skip rows with dirt.

"You're covering them too deep," Billy said, then turned to Zach and said, "You're spreading them too thick."

We smiled. This is why Zach makes the perfect farmhand; even when we do exactly as told, it looks like we've still got a lot to learn.

ꝏ

In 2012, Billy's neighbors decided to sell their home. Concerned about who might buy the house adjoining Billy's property, he and I fretted. Powerless and worried, we prayed that the new neighbor would understand, and embrace, Billy's lifestyle. Not everyone enjoys bleating goats and cackling hens. Some people see them as nuisances.

Enter a new friend who traveled to Georgia from the Wild, Wild West. While many toured the home and turned up their noses at Billy's garden and the unobstructed view of his barn, Neighbor Joe and his wife, Helen, saw something special.

"That Joe shore is a good neighbor," Billy often says. "Do you know he's got that Chevy of mine running like a new one."

Watch out, Tinker. It looks like your days are numbered.

"Yup. Ole Joe and me have big plans for the garden. We're going to haul us a couple loads of horse manure to his place, then a couple loads to mine. He's decided to grow some greens at his place. I'll help him with that, and he'll help me with what I need around here. Yup. I shore am lucky to have me a neighbor like Joe."

Joe is not only a fixer but also watches over Billy like a poppa bear. Something Zach would soon learn.

Zach was in the pasture petting kid goats, an activity he had done many times before. Suddenly, he heard rustling in the bushes. Believing it was a raccoon, Zach ignored the noise and continued playing with the animals. Eventually returning to the carport, he encountered a man deep in conversation with Billy.

"Ole Joe here was just asking me about you," Billy said. "He thought you might have been up to no good, said he was checking you out from behind the bushes. You might want to be careful sneaking up on Joe. He used to be in the military."

Frightened, Zach apologized. Then the three men shared a laugh.

"That Joe takes good care of me," Billy shared. "If he hasn't heard from me each morning, he comes a knockin' on my door. I've got to make sure I let him know my whereabouts or he might call the law to check on me."

The law is familiar with the Albertson abode. One of Billy's friends reported him missing several years ago, but it was actually a false alarm. He was visiting friends. After the police became acquainted with him, Billy became a poppa to many of the boys in blue. Several officers frequent the farm. They come bearing breakfast. They bring their friends and their children. They write words scribbled on tattered pieces of paper. *Stopped by to check on you. Will return later.* The City's Finest go above and beyond to create a relationship with all they serve and protect; for that, we are grateful.

In case you are wondering, Billy's daughters haven't asked Neighbor Joe to take care of Billy. Neither have I. A still, small voice whispered to Joe's heart, and he listened. A good neighbor looks in on his fellow man. I am happy that Billy now has someone next door whom he can call should he need help. I am also thrilled that many people, regardless of their geographic location, stop by and check on him. His daughters and I hope all of Billy's friends continue to be in his life for a good long time.

Friendship Cake

This is one of Billy's favorite cakes. Then again, he loves all cake.

Ingredients
1 box yellow cake mix
1 can (14 oz) sweetened condensed milk
1 can cream of coconut
2 tablespoons rum flavoring
1 (16 oz) can crushed pineapple, with juice
1 pint frozen whipped topping
12 ounces coconut

Bake cake according to directions in a sheet cake pan. Carefully punch holes in cake while it is still warm. Mix condensed milk, cream of coconut, and rum flavoring. Pour over warm cake. Carefully spoon pineapple and the juice over the cake. Frost with whipped topping and garnish with coconut. Can be made a day ahead.

Friendship Tea

There are two types of friendship tea: the powdered variety listed here, which makes an excellent gift, and one shared when friends gather. The recipe for both follows.

Ingredients
½ cup instant powdered tea (unsweetened)
1 cup sweetened powdered lemonade
1 cup Tang
1 teaspoon ground cinnamon
½ teaspoon ground cloves
Pint jars

In a large bowl, add ingredients and stir gently to mix. Pour carefully into pint jars and adorn with colorful labels.

To serve, spoon two teaspoons of powder into a mug of hot water.

Mrs. Redmond's Friendship Tea

Mrs. Redmond shared this recipe during the wake of Aunt Jackie Rountree. One of Jackie's favorite beverages, it is the perfect super-sweet tea for social gatherings.

Ingredients
2 quarts water
2 small containers frozen orange juice
½ cup lemon juice
4 family-size tea bags
2 cups water
2 cups sugar

Place 2 quarts of water in a saucepan and heat until boiling. Add teabags and boil for 5 minutes.

In a separate saucepan, boil 2 cups of water and add 2 cups of sugar.

Pour cooled sugar water into a large pitcher. Add frozen orange juice and lemon juice. Add strong tea. Mix well. Enjoy with friends.

Friendship Cookies

Delicious cookies to include with any recipe swap party.

Ingredients
1 cup vegetable shortening
1 cup granulated cane sugar
1 cup brown sugar
2 eggs
1 cup evaporated milk
1 teaspoon baking soda
2 teaspoons baking powder
½ teaspoon salt
1 teaspoon vanilla
1 tablespoon white vinegar
4 cups plain, all-purpose flour

You will need to roll cookies in sugar before baking. For that reason, you will need an additional bowl of each of the following:
½ cup granulated cane sugar
½ cup confectioner's sugar

Preheat oven to 375 degrees.

Cream shortening with brown and cane sugars. Add eggs. Mix well. Add baking soda, baking powder, salt, and vanilla. Mix well. Add 2 cups of flour, then the evaporated milk and vinegar. Add remaining 2 cups of flour until everything is combined.

Drop dough by teaspoons into powdered sugar, then transfer to the bowl of granulated sugar and roll to coat. Place on cookie sheet and bake for 10-12 minutes until lightly brown.

16

Failing to Communicate

Communicating effectively is crucial to all relationships. Usually, I am a good listener who understands clearly spoken words, but when canning season rolls around, my ability to focus becomes limited to cooking dinner, doing the laundry, and running the pressure cooker nonstop.

The Ace tomato is a new variety we planted at Billy's. It is perfect for those who like to preserve tomatoes and make spaghetti sauce. This particular variety also provides a high-yield harvest.

I was working as fast as humanly possible, smiling as I separated the inside from the peel, enjoying the slippery satisfaction as I pinched the peel and dropped the juicy fruit into sparkling clean jars. Lines of perfect tomatoes filled the countertop. To my left, a bushel basket promised a long day of peeling and processing. Though the kitchen looked messy, being surrounded by it all rewarded my hard work.

My canning method is simple. First, I sort the tomatoes according to size. My favorites are Ace, Brandywine, Black Krim, and Cherokee Purple. I dehydrate the largest tomatoes, then can the smaller tomatoes whole or use them for salsa. After sorting the harvest that particular day, I realized my food dehydrator was still at Billy's. I had loaned it to Janet, who had fallen in love with dehydrated tomatoes after nibbling on a desiccated Black Krim.

Abandoning the mess, I arrived at Billy's and was immediately greeted with a Mac Attack, an onslaught of Kelle Mac's boys, who rushed to say hello and sandwich me with smiles and hugs so love-filled I could barely contain the joy. Their father, Todd, had formed a shelling circle. Each boy perched on a bucket, separating peas from their husks. Their fingers dropped peas into a bowl positioned in the center of the circle.

Pea shelling is the perfect project for young children. Nimble fingers easily separate pea from pod at a pace much faster than adults can maintain. Billy barely has time to harvest the peas, much less unzip them for later use.

After I retrieved my dehydrator and explained the contraption, I asked, "Did Kelle ever get a food dehydrator?"

Shaking his head, Todd answered, "I don't think so."

"I'm also canning salsa," I volunteered.

This statement peaked Todd's interest.

It is common for those who preserve their own food to exchange recipes with anyone who shows even a moderate interest. After explaining how easily one can prepare delicious salsa, I said, "If you guys eat a lot of salsa, I can come over and show Kelle how to make it."

Then I *thought* Todd said, "Sounds great. I like mine spicy with lots of peppers. We'll be gone next week, but you can show her how to make it when we get back."

Kelle despises wasting food. When the family leaves town, she issues a come-and-get-it open invitation for her friends to step into her garden and harvest anything they want. Knowing this, I said, "I'll just pick while you're gone and convert your tomatoes into salsa."

Todd agreed.

Since I knew Kelle would be exhausted from her trip, I envisioned her joy upon seeing beautiful jars lovingly displayed in her kitchen.

That was my best-laid plan, filled with good intentions. Aren't they all?

Tuesday: I sent Kelle a text that read, *I'll make you some salsa today.*

Her reply: *I'll trade you a cherry pie.*

Yes, I am texting now. A reluctant push into the twenty-first century dictated by the pressure I experience in this instant society. Apparently, I *was* the only person in the universe without texting. I hate texting, with its impersonal communication and the abbreviated I'm-too-busy-for-you, made-up words. Most people welcome a phone with touch-screen texting. I don't. Primarily because, without fail, every time I placed the phone in my back pocket, my posterior dialed people in my contact list. The youth of today call this process "butt dialing." I also send the occasional text message with my rear end and capture images from inside my pocketbook. Remarkably, I cannot figure out how to use the camera when I need to take a photo.

Tuesday passed quickly. I was unable to work through all of my own tomatoes. *No worries, I'll just pick Kelle's on Wednesday while I'm running errands.*

Wednesday: I arrived at Kelle's and found her patio table loaded with tomatoes that were sorted according to size and ripeness. The assortment of tomatoes stacked with military efficiency surprised me. I thought, *Wow. Whoever is caring for her garden is doing a great job.* I foraged through the peppers, picking just enough to make a run of salsa, then returned to the table carrying my empty basket. As I carefully placed tomatoes in the basket, my conscience whispered, *You might want to call Kelle and make certain she is out of town.* To ease this concern, I dialed her number. Her pre-recorded voice encouraged me to leave a message.

"Hi, I'm checking to make certain you are out of town," I said. "You have a bunch of tomatoes on the table. I'm taking all of them for salsa. See you soon. Love ya."

Imagining that she was lounging at the pool enjoying vacation time with her boys, I still felt a hint of uneasiness. I followed the voicemail with a text: *Geez. I hope you are out of town. If not, I took all of your red tomatoes on your table.*

One of the reasons I deplore texting is that people seem to bypass verbal communication, preferring the transmittal of cryptic messages that are easily misunderstood. Kelle's return text was direct, leaving no room for doubt.

Girl. I'm making sauce with those!!!

Three exclamation marks equals Zippy was in big, big trouble.

My response: *Not anymore. I just put the salsa on the stove to cook :).*

The insertion of a colon and a parenthesis is a whimsical punctuation tool used in texts when one is too afraid to speak face to face. It's the code for a smiley face. Think of it as a way of easing the news that I had just sashayed onto her property and loaded up two bushels of tomatoes.

Kelle's response: *So does that mean I actually have to buy maters for mater sammies tonite?*

My response: *Uuuhhhhh. Seriously. I thought you were out of town.*

Aaaaakkk!

With texting, the use of consonants and vowels consecutively in a word or phrase symbolizes intense excitement or, in my case, distress.

Even with deep friendships, mistakes happen. My intent to ease Kelle's tomato burden had taken a horrible turn. The amount I had stolen…I mean used…was small when compared to the tomatoes waiting for processing. Abandoning the pressure cooker, I loaded the tomatoes back into wicker baskets and prayed Kelle would not be home when I returned them.

A serious gardener may let you take her husband, perhaps even her children, but steal tomatoes from her back porch in the broad-open day-light, and she's liable to smack the freckles clean off your face. Swiping maters is grounds for some serious fightin'. Because my friends and I are always joking around, and because I was still uncertain how I could have possibly misunderstood the original discussion with Todd about Kelle being out of town, part of me still believed she was joking until she sent the following text: *If you have any of the tomatoes left, I really was planning on using them. I've been purposely picking them for sauce and sammies. We don't leave town until next week.*

Next week? *Aaaaakkk!*

As soon as the pressure gage displayed zero pounds, I transferred the piping hot jars of Mac Attack Salsa into a box and loaded them into the car. Fortunately, Kelle was still not home. Feeling the gaze of nosy neighbors, I returned everything, trying my best to display the tomatoes just as I had found them.

There. Good as new. You can't even tell I've been here.

I hoped.

I also hoped Kelle would like the salsa enough to still be my friend.

At least she knows my heart. Since I used the smallest tomatoes first, she had plenty of the larger ones for sammies. The jars did look beautiful sitting on her kitchen counter.

Mac Attack Salsa

This salsa recipe is for those who want to preserve a large amount for later use. Readers interested in canning their own food should purchase a copy of the Ball Blue Book. I prefer the older editions because they contain more traditional recipes. My particular book is the 1966 edition.

Ingredients
4 cups tomatoes, chopped
2 cups onions, chopped
2 cups green peppers, seeded and chopped
3 jalapeno peppers, seeded and chopped
2 cups white vinegar (must have 5% acidity)
1 ½ teaspoon salt
1 tablespoon minced garlic
1 tablespoon fresh cilantro
1 tablespoon oregano leaves
Pint jars with rings

Wash all vegetables thoroughly. Combine ingredients in a large saucepan and bring to a boil. Stir frequently. Simmer 20 minutes. Ladle hot salsa into pint jars. Process in boiling water bath for 15 minutes.

Zipper Peas

Zipper peas are a staple in the Albertson home. Billy enjoys them year round with a large onion for added flavor.

Ingredients
2 cups fresh zipper peas
2 cups water
1 medium onion, chopped
Pieces of country ham or seasoning meat
Pepper to taste

Rinse peas and place in saucepan with 2 cups of water. Bring to a boil and add several small pieces of country ham or seasoning meat.

Boil on medium for 45 minutes.

Add chopped onion and boil for another 10 minutes. Add a pinch of pepper.

Test a few peas to see if they have reached the desired consistency. Serve warm.

Spaghetti Sauce

This recipe is for those who have enough tomatoes to can large quantities of sauce. While tomatoes are acidic, the addition of vegetables in the sauce reduces the acidity. For that reason, this sauce must be either processed in a pressure cooker or frozen immediately.

Ingredients

½ bushel of tomatoes, chopped
8 medium onions, chopped
6 bell peppers, chopped
¼ cup vegetable oil
½ cup sugar
¼ cup salt
42 ounces canned tomato paste
2 tablespoons dried oregano
2 tablespoons dried basil
2 tablespoons dried garlic

Wash all vegetables. In a large saucepan, add vegetable oil and chopped vegetables. Bring to a boil, then reduce heat until vegetables begin to simmer. Add sugar, salt, tomato paste, and spices. Cook for an hour or until sauce has reached desired consistency.

Processing the Sauce

Pour sauce into jars. Seal and process in a pressure cooker using this guide:

Pints—15 minutes at 10 pounds of pressure

Quarts—20 minutes at 10 pounds of pressure

This sauce can be frozen using freezer bags and will keep for approximately one year in the freezer.

17

Milking Goats

On the farm, the end of winter brings the promise of longer days, warmer weather, and the arrival of baby goats. The kids come in pairs, and sometimes as triplets who are rarely identical but always precious. The "mothers," as Billy calls them, gently nudge their kids to an udder heavy with warm, rich milk. Birthed with a voracious appetite, newborns press hard against their mothers, tails wagging in excitement, as milk flows into their fuzzy pink mouths.

Billy often gets requests for goat milk from parents whose children are allergic to cow's milk. He raises pygmy goats, a low-to-the-ground breed instead of the larger, milk-producing Nubian.

For years I had hinted, begged, hoped, and finally given up on learning how to milk. Then one day the opportunity arrived. My dear friends Donna Baker and Ana Raquel visited the farm. During a stroll through the property, we stepped into a field of yellow buttercups and I noticed a nanny whose udder was swollen and distended.

Searching the pasture, I asked Billy, "Where is her baby?"

"Oh, I sold her kid this morning," he said casually, then quickly added, "but it looks like I need to relieve her."

I nodded, knowing that the term "relieve" meant the manual extraction of milk.

Finally, my goat-milking chance had arrived. But when Ana said, "Oh, Mr. Billy, can I please milk the goat?" I couldn't be rude to a guest and whine that I had waited years for this moment.

I did whine to Ana, later…in jest.

Upon seeing the goats, Ana, who was visiting the farm for the first time, immediately shared stories of growing up in Puerto Rico, a place where her father still lives. How could I steal the chance for her to touch a childhood memory and recall a personal snapshot of home? I couldn't take away the opportunity for her to call her father—as I often called mine—with a Farmer Billy story.

Milking goats at Billy's is a team effort. It is impossible to coax a nanny onto the ramshackle platform. One person pushes, another pulls, and a third shakes the feed bucket. Despite this method, once the nanny places a foot on the wobbly stand, the boards creak and sway, which triggers her flight mode. Miss Nanny begins looking for an exit. Often the only means of escape is through the person nervously shaking the feed bucket while backing out of the way.

By my estimation, the milking stand is at least twenty years old. Constructed of untreated lumber, it weathers the elements and, most often, serves as a coat rack and bucket holder. Billy runs the lawnmower around the structure during the summer and completely ignores it in the winter. He does expect the contraption to work whenever called upon, despite the lack of maintenance. Judging by the leaning structure, a strong puff of wind could topple it at any moment.

Yet at the mention of milking, Billy smiled at Ana and said, "Why, of course you can milk this here goat." Turning to me he said, "Zippy, fetch a vessel while I get something to tie the momma with."

He plucked a string hanging from a nail inside the barn. Frayed fibers fell to the ground as he knotted them into a makeshift halter. Then Ana, Billy, and I wrestled the nanny into the rickety guillotine-shaped vice that only partially secured her. Scurrying to the front, I offered the feed and prayed for a speedy milking experience. Donna captured the moment on camera.

"Here," Billy said while placing a cup in Ana's hand, "let me get her started."

With expert fingers, Billy reached beneath the nanny and squeezed. Moments later, he said, "Zippy, fetch me a few more vessels. She's just full of milk."

Passing the bucket to Donna, I dashed to the house.

Acquiring vessels from the Albertson kitchen is a dangerous task. Despite multiple attempts by myself, and others, to sort and stack reusable plastic containers in an orderly fashion, those familiar with Billy's lifestyle know to cautiously open a cabinet with one hand while shielding themselves from falling objects with the other. I haven't determined exactly where all the pieces go once I leave his home; I just know that when I opened the cabinet, the only thing I could find was a Cool Whip container, a 32-ounce convenience store cup, and a couple hundred assorted lids that were missing the matching bottom section.

When I finally found a container and went back to the milking stand, Ana took a seat near the nanny. In a matter of seconds, milk splashed into the plastic container Billy held. She was an expert. As warm drops of milk filled the cup, Ana looked at Billy with a face that revealed pure joy.

A bond forms during visits to Billy's. Threads of just-made memories weave their way into a special place in the heart. Images captured on film, smiles shared, and that magical moment when a baby goat nibbled your pants or a chick hid beneath your hair: these experiences nestle deep inside a lonely place we all carry, a place we do not share with others. Visits to the farm are precious, priceless, and often recalled in unexpected moments during ordinary days. Sunlight pierces the cloud. A breeze kisses our face, and we recall that moment when old friends introduced us to new friends. In that moment, we became family.

ঞ

The first wobbly moments of a baby goat's life are so precious that even a farmer who has witnessed the bonding encounter many times can't help but stop and smile.

"Watch that little one wag his tail," Billy said as we leaned over the fence, marveling at the newness of life. "Isn't that something? That Silver is a good Momma."

But not all mothers possess maternal instincts.

Ash Wednesday, 5:45 pm

As Billy and I visited inside, Jamie slipped around the back of the house and into the pasture. We were chatting about spring fever when my cell phone rang.

"You've got to come to the barn," Jamie said. "There is a baby without a momma."

Most visitors to the farm would not realize the wisdom, or urgency, of her call. Jamie understood that newborns nurse every few minutes and rarely leave their momma's side, especially when there are so many other larger goats around.

Rushing to the barn, I met Jamie, who was cradling a skinny kid, its frantic yelp sounding exactly like, "Momma. Momma."

Except no momma came.

As I went out and quickly scanned the field, it appeared as though every female had a baby or two. In the distance, a small goat meekly approached Jamie, responding with a gentle call. Each mother-baby pair has a unique call that identifies them in the pasture. Baby goats sometimes get lost in the grass when they scamper off to romp with other newborns. When they lose sight of their mommas, they panic and belt out a few notes. She responds in kind. After rushing to their mommas for a drink and a gentle touch, the kids return to their playmates. However, this baby was different. He toddled behind his mother bleating, crying. His hungry calls went unanswered. She didn't allow him to nurse. She ignored. She shunned. She abandoned.

When I met back up with Jamie, she and I exchanged a worried look. "Take the baby to the stall behind the barn," I instructed. "Maybe the momma will follow."

Cries filled the air as the kid struggled to escape the bonds of human hands. After capturing the nanny, which was quite a rodeo, Billy secured her to the fence. Jamie held the baby, and I had the privilege of finally learning how to milk.

The wailing of a hungry goat is a sorrowful thing. This one's incessant cries might lead the neighbors to believe we were offering him in a ritualistic sacrifice. Jamie explained that she had named him "Ashton" because he was born on Ash Wednesday, and it looked like his personality was as somber as his birthday. Both he and his mother wailed as we crowded around. Jamie held the newborn, and Billy relieved the mother of milk while I attempted to open the kid's mouth for him to either grab a teat or capture the milk Billy shot in his direction.

Their cries of help were justified.

"C'mon, little feller, give us a hand," Billy encouraged.

Everyone was frustrated. Jamie did not understand why our efforts were not working; neither did I. Billy, ever calm in this type of situation, continued to try. Taking turns soothing the first-time mother, we stroked her belly.

"You're a good momma," Billy murmured. "You just need a little help, that's all."

Insert into this experience the three guests who suddenly dropped by. I hoped they understood that we could not entertain them. Visitors usually receive hugs and garden tours, but on that evening, nighttime was approaching, and this matter was of the utmost importance.

"Can't Billy just feed it with a bottle?" one of the women asked while leaning across the fence.

Shaking my head no, I explained that bottle-feeding is a short-term solution. Bonding the pair is much more important. Besides, Billy doesn't have a bottle on the farm, and I didn't have enough daylight left to go out and buy one. Then Billy announced, "I had to bury her other baby this morning. It was too weak to break out of the birth sac."

This new information spurred me into action. There was no time to chitchat with visitors. My sole focus was getting the newborn to nurse. Despite our attempts, the mother's udder was so heavy that little Ashton could not get the swollen nipple into his mouth. Newborn kids must ingest colostrum within the first few hours in order to stimulate their digestive tracts. On the farm, a sickly mother quickly becomes a critical situation. Mastitis, an infection occurring in a doe's udder that prevents her from making milk, is life threatening to both the mother and her kid.

Billy covered my finger with milk, which I placed in Ashton's mouth, then coaxed him beneath his momma. Still, he would not nurse. After filling a container with milk, we tried pouring the liquid into his mouth. That was also disastrous. Frustrated, I made a mental note to bring a baby bottle to the farm.

"Let me give it a try," Billy said while reaching for Ashton. "I'll hold him, you milk."

The nanny's udder was rock hard and warm, a sign that mastitis was encroaching. Copying the technique I had witnessed, I gently wrapped my fingers around her udder and squeezed. No milk came.

"Can you tell me how to do this again?" I asked.

"You need to grab hold and push up a bit. Squeeze from the top near the bag."

Following his instructions, I pushed my hand against the udder, clasped her teat with my thumb and forefinger, and squeezed from the top down. Miraculously, a stream of milk shot forth, bypassed the container Billy held, and struck my cheek.

Droplets fell from my hair into the grass. Billy and I laughed.

"How in the world do you ever hit the cup?" I asked.

"You just have to keep practicing."

Concentrating on my aim, I tried again and again. Where was Ana? We needed her expertise. When I finished, milk was everywhere, except inside Ashton's tummy. Droplets formed a goatee across his furry face. Tiny pieces of grass adhered to the milk that was gluing my fingers together. Spots dappled Billy's glasses. Still, as darkness fell there were no signs that the little one understood how to nurse. He only knew that he was very hungry. We only knew the frustration of failure.

Thursday Morning, 8:00 am

My first thought the next morning was Ashton. Had he survived the night? If so, could I convince him to eat? After purchasing a bottle for premature (human) infants, I climbed over the goat fence while Billy, who struggles with insomnia, slumbered. Early morning visits are tiptoe quiet. Even Billy Albertson deserves his beauty sleep.

Entering the pasture at first light without a feed bucket is a dangerous task. The sudden rush of nannies and their kids left me feeling claustrophobic. I prefer feeding from safely outside the fence. Billy's goats are not aggressive, but they *are* demanding, especially in the early-morning hour. Tucking the bottle into my jeans, I displayed empty hands while repeating, "I don't have any food. I'm just here to check on the baby." The nannies followed me to the barn.

Billy had secured the stall with a variety of strings, wires, and pieces of scrap lumber, creating a hodgepodge of hazards for any human, or goat, who wished to gain entrance. The barn isn't wired for electricity. Without a light, I fumbled for a handle, lever, something…anything that would allow me to open the stall. Finding none, I saw that the only way into the barn was over the mess. Transferring the baby bottle from my waistband to my mouth, I shimmied over the gate and prayed while my eyes adjusted to the darkness inside the stall.

Ashton was alive.

Jamie hadn't been worried. This morning she was confident, without a smidgen of doubt that Ashton had survived. Unlike her mother who fretted and prayed. Mothers often tell their children that everything is going to be fine while secretly praying that their words will ring true.

Ashton, like most babies, was curious. He greeted me with a pink-lipped smile and then cautiously inched forward. Crouching on all fours, I extended my hand and sat still while he approached. A soft nicker escaped from his mother, reminding him of last night's torture. After gently placing him in my lap, I guided the bottle toward his mouth. I reasoned that getting some water down him was beneficial. As expected, he fought the bottle, drenching us both in the process.

Fortunately, his mother didn't charge. She remained in the corner. As I released him, I attempted an examination. Tilting my head and bending low, I observed that her udder appeared empty. Ashton bounced over to his mother and hid behind her.

He did not nurse.

But he appeared noticeably more energetic, with a tummy that felt fuller. Perhaps they had bonded during the night.

Crawling beneath the manger, I enjoyed a moment of peace. Rough-cut boards, with nails bent and hammered flat to prevent injury, lined the walls and supported my back. Hay provided warm bedding. Buckets of food and water sat in the corner. Sitting in the straw among the goat droppings, I marveled at this wondrous life, this gift, this promise of new beginnings.

An hour passed before Billy appeared. He reached a hand inside the door, lifted a lever from the inside, and entered the stall without the need for shimmying.

Noticing me in the darkened corner, he said, "Well, well. What in the world are you doing here?"

"The little one was on my mind first thing this morning," I explained. "I'm just checking on him. Right now, I can't tell if he has nursed."

"Well now, he looks quite pert. I think he's gonna be okay."

Glancing back at Ashton, I agreed. He was definitely livelier than he'd been the night before. "Do you think we should keep them separated from the older goats until they figure things out?"

Billy agreed, then added, "Let me secure the gate."

He disappeared for a moment, and the clang of metal and the rattle of chain signaled that it was safe to release them from the barn.

"It's a beautiful sunny day," Billy said with a glance heavenward. "A little sunshine will do 'em both good." He opened the door.

Ashton followed his mother to the edge of the pasture. While she nibbled on new grass, he bumped her belly. Then his mother looked directly at me and nickered as he opened pink lips and nursed.

Billy draped an arm around my shoulder. "See there. Everything's gonna be okay."

Homemade Ice Cream

Do you remember enjoying hand-cranked homemade ice cream? Mmm, I do. With the advent of electric ice cream makers, we can now enjoy homemade ice cream as often as we want. Give this recipe a try for a true taste of summer.

Ingredients
1 can of sweetened condensed milk
1 (12 oz) can evaporated milk
2 tablespoons real vanilla extract
1 pinch salt
½ cup pure cane sugar
6 cups milk
1 bag of ice
1 box ice cream salt

Mix liquid ingredients and store in refrigerator until ready to use. Pour ice cream mixture into a maker, then follow the machine's instructions. Pour a layer of ice, then sprinkle a layer of rock salt over ice. Repeat until churn is surrounded with ice. Note: If you would like to make a lower-fat version of this ice cream, substitute coconut milk for the evaporated milk. You can also use low-fat milk, but the ice cream flavor will not be as rich.

Goat Cheese

It takes a lot of milk to make goat cheese. This recipe is for the overachievers.

Ingredients
4 quarts goat's milk
1 pinch salt
⅓ cup distilled white vinegar
½ teaspoon dried minced onions

In a large saucepan, bring the milk to a slow boil over medium heat. Be careful not to let the milk scorch. Add dried onion and salt. As the milk begins to bubble, turn off heat and pour vinegar into the milk. The addition of vinegar causes the milk to curdle. Once that happens, pour it through a cheesecloth-lined colander. Shape into a ball and keep refrigerated until ready to serve.

18

Jill

Babies sometimes need extra care during the first weeks of life. This was the case with another goat we called Jill. When admiring Jill from a distance, I noticed she struggled to maintain her balance. While her hooves appeared perfect, upon further inspection I discovered that they were stubbornly curled, fist-like, a defective anomaly that forced her to walk on what I called her "knuckles."

Even though she appeared to suffer no physical discomfort, every movement forced her hips and hind legs high above her shoulders.

"You should have the vet drop by and check her out," I suggested.

Billy has a long-standing policy of letting Mother Nature take her course. So when he replied, "She'll figure it out," I doubted the vet would make an appearance anytime soon.

Watching the tiny black and white goat hobble in the field hurt my heart. She wanted to scamper and jump like the other kids. She tried but failed. Tears pricked my eyes. After opening the gate, I scooped her into my arms and gently uncurled one hoof. When she didn't protest, I worked the other. Placing her on the ground while carefully supporting her weight, I groaned when she couldn't support her tiny frame.

"She'll manage," Billy said, confident that her condition was temporary.

Billy probably secretly wishes his girls didn't get emotionally attached to animals. From the tiniest chick to the largest goat, Jamie and I pray for the sick and become angry at the first sign of barnyard bullying. We are surrogate mommas to all God's creatures at the farm.

While Billy went about doing other outside chores, I worried that the crippled little goat would not manage. In my heart, I believed the others would single her out as different and shun her or, worse, began physically pushing her around.

Morning dawned, and with it came a series of physical therapy treatments for Jill. With no formal veterinarian training, I uncurled each hoof and then pressed her weight against my palm.

I like to believe that she understood I was trying to help.

Within days, her condition improved, and we bonded. She waited for me at the gate and welcomed me with a series of calls. She also developed a fondness for the hem of my jeans. She often grabbed the fabric in her mouth and pulled as I walked across the pasture.

"The more you handle goats," Billy says, "the easier they are to handle."

It is true; goats that interact with humans at an early age tend to bond with them. They also separate from the herd and prefer the company of their human friends. Jill developed such an easy-going persona that she was invited into the kindergarten classrooms of many public and private schools. After loading her into a large crate, Billy carried her to school, where the students fell head-over-heart in love. Like most parents who take their children to school, Billy did not stay; he left Jill there until classes were over. Then a parent returned her at the end of the school day.

Her charm and loveable personality taught children about animals and encouraged adults to embrace diversity. Goats have a reputation for smelling bad and eating anything placed in their path. In Jill's defense, if you washed your hair with strawberry shampoo, you should have expected her to nibble. She just couldn't help herself.

Jill was so accustomed to human contact that she called out any time someone approached the pasture. Her personality also made her highly desirable to those who stopped by the farm in the market for a goat of their own.

For the record, Jill was not for sale.

Billy had recently expanded his garden by moving the fence and planting beans alongside the goat pasture. While harvesting beans beneath the vines, I overheard a customer express an interest in Jill. I don't usually eavesdrop on conversations, but at the sound of Jill's name, my ears warned of trouble. But then someone said, "I really like that small black goat. What will ya take for her?" It was panic time.

Crouching behind the foliage, I spied the beginning of a disastrous deal. Opening my cell phone, I placed a call.

"Janet," I whispered to Billy's daughter, "we've got a problem."

Janet has heard these words countless times before. Sometimes I've called because Billy was pushing himself too hard, the refrigerator was leaking again, or he had almost burned the house down during a do-it-yourself-project. This time the problem was far greater than Billy rewiring the stovetop. This was an actual emergency.

"Someone wants to buy Jill," I whispered.

"Well, they can't have her," Janet immediately replied. Her voice boomed through the tiny device.

Taking a seat on an empty five-gallon bucket, I said, "Billy sounds pretty serious. I can't just barge over there and tell one of his customers that she can't have Jill. They'll both be mad at me. What are we going to do?"

"Let me think for a minute," she replied.

Janet was a quick-on-her-feet thinker. She was the perfect gal to call in any kind of emergency, especially one involving the sale of our beloved Jill.

Peeping from behind the broad leaves, I asked, "What are we going to do?"

I waited anxiously while Janet formed a plan.

Billy had recently purchased what he called a "cellular device" from Walmart. The device was safely tucked inside a plastic bag and zipped inside the front pocket of his bib overalls.

"I'll have Kristen call his cell phone and say she is looking forward to playing with Jill this weekend. That will buy us some time."

Moments later, Billy unzipped his overall pocket, opened the bag, pressed the phone to his ear, and said, "Yello."

Operation Rescue Jill was a success.

Believing that she understood their words, children and adults alike wrapped their arms around Jill's neck, whispered secrets into her ear, and sealed the bond with a tight squeeze. Jill responded with a gentle nicker and a couple flicks of her tail. This was her equivalent to a "pinky promise."

It seems that Jill, like some people, was destined to struggle with episodes of ill health. When Jill was in the motherly way, her pregnancy caused Billy and I much concern. She grew larger with each passing day until she reached the point where she could barely walk.

"I think you need to call Denise's friend, the vet, to come over and check her out. She may need a C-section."

Billy didn't have time. Before he placed the call, he discovered Jill in the midst of labor, with both she and her unborn kid in severe distress. The kid was too large, Jill too inexperienced. Several hours passed with both Jill and Billy exhausted by the birthing process. The baby goat did not survive. After a proper burial, Billy returned his attention to Jill, who was weak and still in distress. Denise's friend administered antibiotics and left a series of shots with Billy to stave off a life-threatening infection. Jill wouldn't be able to have any more babies.

Like human mothers who lose their children, Jill mourned her baby. Soon after gaining enough strength to walk, she searched the field for her offspring, stopping only long enough to collapse with fatigue. Her heart-broken cries filled the air.

"She misses her baby," Billy explained as he and I dropped corn seeds into the earth.

"Oh, she is searching for her baby. This just breaks my heart," I replied. "Poor Jill."

Then Jill did something so precious that Billy and I stopped working. She approached a goat that was not hers and allowed him to nurse.

"Would you look at that?" Billy whispered.

Tears stung my eyes as Jill nudged the light brown goat toward her swollen udder and encouraged him to nurse. Billy and I both smiled, understanding more than ever how special Jill was. For many weeks, both Jill and the kid's real mother fed him until he finally determined to leave his birth mother and stay with Jill.

ꟹ

Each of my visits to the farm begin with a "Ho there" greeting. After Billy's greeting one day, I stepped behind the chicken lot and scanned the field for Jill's familiar black and white face. One day I couldn't locate her. I imagined that once again she was on loan to one of the local private schools in the area.

"Where's Jill?" I asked.

I was shocked when Billy said, "I sold her."

Oh, my heart.

"What? Who did you sell her to?" I demanded.

Billy shrugged his shoulders, "I don't know. Some lady came by and bought Jill's adopted son. I didn't want to sell Jill, but she made me an offer I couldn't refuse."

Trying to remain calm, I asked, "How much?"

Then Billy, who is the most honest individual on the planet said, "Well, she hasn't exactly paid me for Jill. She only had enough money for the kid. I let her take Jill with her. She said she'd come back and pay me in two weeks."

"When was that?"

Billy delayed his response before saying, "It's been about two weeks, maybe more."

Feeling a mixture of disbelief and heartbreak, I asked again, "Who is she?"

Shrugging his shoulders, he responded, "I dunno. Some lady who owns a miniature horse farm up in Milton."

"Where?" I demanded.

Another shrug.

I adore this Billy Albertson behavior—always trusting someone to do the right thing. Still, I was skeptical. Two weeks is plenty of time to pull the money together.

Abandoning the weeding and other chores that waited, I rushed home to call Janet. Tears streamed down my face as I gulped huge pockets of air. Billy would not understand my pain. He doesn't get attached to animals. To him, they aren't pets or family members. But I have always confided in animals. They are the only confidants who keep my whispered words tucked safely inside their hearts. Animals never, ever share secrets or break promises.

The year had started with the unexpected death of Mr. Coleman, and then a couple months later Andrew Wordes died. After another few weeks, I learned that a dear friend who frequently visits the farm had a terminal illness and that my mother would need to endure more chemotherapy. It seemed that I was losing all of my friends. While some may mock my friendship with a goat, when I hugged Jill, she understood. She called to me from the field. She missed me when I wasn't there. She demanded affection. She craved my caress as much as I needed her hugs. She knew firsthand about heartache.

"Billy sold Jill," I said to Janet while clutching the phone. She already knew but hadn't the heart to tell me.

"This is his way," she explained. "This is part of living on a farm. Growing up, I learned not to get attached to animals. The cattle were butchered for meat, and the goats were eventually sold."

I cried even harder, hiccupped, and pressed a hand to my broken heart.

"I know you may think I'm being irrational. It's just that everyone I know is leaving me," I explained. "I feel so alone. How could he put a price on Jill?"

Janet sympathized. "I know this is hard. I'm not happy about this either, but Daddy promised me that Jill will live a good long life on this woman's farm."

"Do you know she hasn't paid him yet?" I offered. "I have a problem with that."

I also have a problem with the people who read my first book and have used it as an opportunity to take advantage of Billy. I believed this was the case with Jill. Technically, when someone removes something from a piece of property without paying for it, a law is broken. And while this unnamed woman may not have meant to steal Jill, as the days accumulated into weeks, my optimism of her honesty solidified into the cold, hard truth. She did not intend to pay for Jill.

"Tell me again where this woman was from?" I prodded weeks later.

Wise Billy either knew exactly where she lived or knew that if he told me, I would drive straight to the woman's home, buckle Jill into my passenger seat, and return her to the farm where the children who loved her could visit as often as they want. Or he may have determined to let bygones be bygones and let Jill go without receiving payment. This is why I called Sweetheart.

Billy's youngest daughter, whom he calls Sweetheart, and I are almost twins. With only a few days separating us, we believe in fairness. Rules are rules, especially when it applies to Billy. Sweetheart also has a reputation for setting unruly customers on the straight path in a snap.

"You know it's been over a month and a half since she took Jill," I said to Sweetheart, then added, "I've identified an area and have people searching for any sight of her. I think I have determined where Jill is."

"I'm going over to Daddy's today and see if he will give me any more information," she replied. "I'll find whoever has her and let you know."

"Good luck," I said. "The only information he will give me is that the woman needed the goats because a film crew was coming to her farm. She needed the place to look like a real farm. I'll keep digging, and you let me know what you find out."

Searching Google, I located a business where I believed Jill now resided. The owner raised both miniature horses and goats. Her website listed a variety of goats available for exorbitant amounts of money. Though a website photo was inconclusive, I believed a recently posted Facebook picture included an image of Jill. Through the process of elimination, I honed in on two possible locations. All I needed was a street name and a trip to the station for a full tank of gas.

Distressed that Jill was trapped in substandard housing, I left Sweetheart a message: "I think I've found her. If she is where I think she is, I am very upset. This particular farm looks like a breeding factory. I'm doing a drive by this week just to be sure. Then I'm going to talk to the policeman who stops by Billy's house."

Enter Billy's other daughter, Janet. Unbeknownst to me, she wanted to participate in Operation Reclaim Jill. Here is how things turned out.

Janet and Sweetheart set out for a road I will not name in the interest of protecting the safety of the property owner. I was not invited. Midway through their mission, they realized they didn't have driving directions. Though I knew the way and would have liked to go along, their gut instinct was correct. Leaving me at home was best. I am not tactful or politically correct when dealing with liars and thieves. Had Andrew Wordes still been alive, he would have rescued Jill a month earlier by driving to the woman's doorstep, loading Jill, and daring the crook to say a word.

Motoring down the two-lane road, the sisters spied a pickup truck pulling out of a driveway and stopped for directions.

"Do you know about a miniature horse farm in this area?" they asked.

I imagine that their pulses quickened when the woman replied, "I own a miniature horse farm."

Inwardly smiling that fate had landed them smack-dab in the driveway of the goat thief, Sweetheart introduced herself and then her sister and ever so politely said, "We're here to pick up Jill. You bought her two months ago from our daddy, Billy Albertson, and haven't paid for her."

The sisters were smug as they watched the color drain from the woman's face and her helpful expression change to one of shock. "I only picked her up two weeks ago and have just been too busy to take him the money," she defended.

"That's not a problem," Sweetheart said. "We are here to collect the money."

Using the excuse that she didn't personally know either of Billy's daughters and didn't feel comfortable giving them cash, the goat thief refused to pay but quickly added, "I'm kind of attached to Jill."

"So are we," Sweetheart replied while walking toward the property. "This is why we will be glad to take her back with us."

At this point, the goat thief panicked. She promised to bring the money the following day. She apologized profusely and waited for Billy's daughters to leave.

Concerned that Billy would be upset at their forwardness, the daughters returned and confessed what they had done.

"Oh, I guess it's all right," he said. "It has been a while since she promised to pay me."

Two months, two days, and approximately seven hours—not that I was counting.

The following day, Janet greeted the goat thief with a "Hi there, remember me?" followed by a wide smile as the thief placed the balance she owed into Billy's hands, plus a little extra for his trouble.

Jill now resides in Milton. Sweetheart and I keep a watchful eye on her living conditions. Should she exhibit the slightest hint of unhappiness, we will load her in the passenger seat and return her to where she belongs.

Apple Pie Flavored Hard Cider

After a hard day of tracking down people who haven't paid their bills, you might need a little nip of "the recipe."

Ingredients
1 gallon apple juice
½ gallon apple cider
Four cinnamon sticks
1 liter Everclear® Alcohol

Combine juice, cider, and cinnamon sticks in a large pan. Heat for approximately one hour until cinnamon flavor blends with juice. Remove from heat. Add Everclear. Enjoy immediately.

Apple Pie

This recipe uses refrigerated piecrusts. My apologies, but sometimes time doesn't allow me to make everything from scratch.

Ingredients
2 refrigerated piecrusts
Glass pie dish
½ cup unsalted butter
3 tablespoons plain, all-purpose flour
¼ cup water
½ cup pure cane sugar
½ cup packed brown sugar
1 teaspoon vanilla
8 Granny Smith apples, peeled, cored, and sliced

Preheat oven to 350 degrees. Place one of the refrigerated piecrusts into the glass pie dish. Reserve the other one for the top.

Slice apples and place in glass bowl.

Melt butter in a saucepan. Stir in the flour to form a paste. Add water and both sugars, and bring to a boil. Remove from heat. Add vanilla.

Arrange apples on top of piecrust. Pour liquid over apples and stir. Note: It is impossible to do this without eating at least one slice.

Place second crust on top, and pinch to seal top and bottom crusts together. Cut small vent holes in pie.

Place glass dish on baking sheet to collect any spillage. Bake at 350 degrees for 35 to 45 minutes or until pie begins to bubble.

raisin'
Cane

19

Raising a Little Cane

"When I was a young boy, Momma and Poppa grew sorghum cane. Sorghum syrup shore is good on a biscuit."

Billy and I meandered through the garden. As always, he was planning, reminiscing, and, as he likes to say, "chewing the fat."

"Yup. Every farmer had a patch of sorghum cane. Back then there wasn't much sugar."

His reference to syrup triggered a memory of my great-grandfather, Lum Winchester. I was fortunate to grow up knowing my great-grandfather, who died when I was seventeen years old. He was a tall man who literally lowered his head to walk beneath the doorframe as he stepped into Grandma Wonderful's kitchen.

Breakfast each morning consisted of fried eggs, homemade biscuits, and a sausage patty washed down with fresh-squeezed cow's milk. A jar of molasses sat in the middle of the table for all to reach. Some people call processed cane juice "sorghum syrup." Where I'm from we call it "molasses." In parts of Appalachia, the process of cooking molasses is called a "stir-off" and is sometimes followed by a taffy pulling.

❦

In western North Carolina, one can observe the process of converting sugar cane to molasses at the homestead farm in the Great Smoky Mountains National Park. Here visitors can watch a mule turn the rollers of an old-timey press while someone feeds stalks of cane into the metal mouth of the device. Because the rollers can only accommodate two to three stalks at a time, extracting juice takes hours.

The cooker is a large rectangular basin four feet wide by twelve feet long and capable of holding hundreds of gallons of juice. The cooker sits atop an outside brick or block masonry oven. Prior to building a fire, someone examines the cooker and determines if the outdoor oven requires repair. The smallest gap or break in the seal between oven and cooker allows smoke to escape and hover above the juice, ruining the flavor. An application of clay quickly resolves any issues, and then it is time to build the fire.

Maintaining the juice at a constant temperature is a crucial and difficult step. Delicious molasses requires a good fire that burns consistently for several hours. Modern cookers boast multiple sections whereby workers can grade the juice according to color and then transfer it from one section to another using a scoop. Juice is heated until it comes to a rolling boil. During this process, several workers hover over the open cooker and, using a metal scoop, scrape the bottom of the container to prevent the liquid from sticking and burning. Someone also uses a metal strainer to remove green foam that floats to the top. This inedible foam is discarded. As the workers move from one section to the next, water evaporates from the juice and it slowly turns from lime green to greenish brown. Hours later, the juice turns a rich amber color, signaling the completion of syrup making.

Stir-offs are communal experiences. Multiple helpers bring the cane from the field, remove the leaves, press the stalks, cook the juice, and bottle the molasses. All of this is hard work, but the taste of thick syrup spooned over a warm biscuit is worth the effort.

ᘓ

Factoring in Billy's love of sorghum syrup and my love of molasses, I seized the opportunity to procure a handful of seeds.

Mother and I often visit the homestead farm. Her people once lived in the Davis-Queen cabin, a home that was dismantled and transported from its original spot in the backwoods. Later, the National Park Service reassembled the structure on the grounds at the Oconaluftee Visitor Center near Cherokee, North Carolina.

One stir-off day at the farm, while discussing our appreciation of the hard work that goes into molasses making, Mother and I both noticed grain heads accumulating in the field. These seeds resemble a corn tassel. They form at the top of the cane stalk and house numerous seeds.

Tucking a handful of dried tops into our pockets, we smiled and left the field. My mother has the ability to grow anything. I secretly hoped some of her gift magically transferred to me.

"Guess what I've got?" I said to Billy later while waving a plastic bag in front of his face.

"Lawd, there's no telling. What you got?"

"I picked us up some cane seeds."

"Sorghum cane seeds?"

I nodded.

Grabbing the bag from my hand, he asked, "Where in the world did you find sorghum seeds?"

After explaining my predisposition to pluck and tuck seeds regardless of where I find them, I confessed, "Picked them up while watching someone make molasses. Some of the stalks still had the grain head on them. Do you think they'll grow?"

Turning on his heels like a youngster, Billy quickly made his way toward the shed while saying, "Let me grab the tiller."

Behind Billy's house rests a strip of land so worthless, so rock-filled, and so nutrient poor that the only crops fit to grow there are crops known to tax the soil. This is where we once grew cotton. We've also planted sunflowers, peanuts, and now sorghum cane in the area. Rationalizing that no amount of organic fortification will convert this sandy, rock-infested soil, we set out to "raise a little cane."

"Now make sure you scatter those seeds real thick," Billy instructed as I stripped the stalk and sprinkled tiny black seeds into the sandy soil. "You gotta plant cane seeds real thick."

While scattering, I thought, *Merciful heavens, if every one of these come up, the cane will be thicker than hair on a dog's back.*

Two weeks later, the dog's back appeared.

Concerned that the stalks didn't have adequate growing room, I asked, "Should I thin these out just a bit?"

"Sure."

For me, thinning baby seedlings is painful. I have a personal relationship with my garden, and uprooting plants makes me feel like a murderer. I believe that we should be responsible gardeners, planting only what we need while saving the rest for later. I also believe that each fragile cane reed has a right to live. This makes determining who should stay and who should die a difficult process. With Mother Nature's unpredictability, I also want a stockpile of emergency seeds.

So that day, I pretended to plant all of the cane seeds but tucked a few in my shirt pocket.

What can I say? I am a seed-hoarding sicko.

Normally, Billy uses excess or thinned crops to make a new row. But we are out of space. Truly, there is not one inch of available space until we plow up the front lawn, which Billy is considering. Currently, Billy pastures most of the goats across the street. In order to make room for feed corn, he cut the large pasture out back in half. As I yanked the thin reeds of sorghum cane from the soil, I hoped the remaining stalks would thrive and tower high above the rooftop. Cane is a tall plant that some might misidentify as corn. For clarification purposes, Billy and I decided it was best to erect a sign that read RAISIN' CANE. His place requires a plethora of signs. City folk don't know cane from shineola.

Taking a few steps back to survey the small parcel, Billy quipped, "Now we're in the syrup-making business."

Since the ratio of juice to syrup is ten to one, meaning it takes ten gallons of cane juice to make one gallon of molasses, neither of us actually believed we would be in the syrup-making business. Best-case scenario, we might grow enough to squeeze out a quarter cup of syrup.

Cane is a zero-maintenance crop. It requires a small application of fertilizer in the spring, and then all one must do is wait until the harvest. The nutritional content of molasses is surprising. Two teaspoons contain twelve percent of the daily requirement of calcium, thirteen percent of iron, and ten percent of potassium. I bet those mountain folk who walked a mule in a circle and fed reeds through a press didn't know that molasses is good for you. Why, it's practically medicinal, like homemade wine.

Hands down, the best way to enjoy sorghum is on a straight-from-the-oven biscuit. As Billy would say, "That's so good it'll make you want to smack your grandmaw."

Buttermilk Biscuits

Most afternoons, my great-grandpa, Lum Winchester, carved out a bit of butter, put it in a saucer, added molasses, and blended the two with a fork, mashing until they became one. He painted a biscuit or hunk of cornbread and then savored every bite. That was his dessert. Give it a try. You'll be glad you did.

Here's a little secret: If you do not have buttermilk, add 1 teaspoon of white vinegar to 1 cup of milk and allow it to sit for five minutes.

Ingredients

2 cups all-purpose flour
½ cup flour (set aside, use to form biscuits)
¾ teaspoon salt
¼ teaspoon baking soda
¼ teaspoon baking powder
4 tablespoons salted butter, frozen
1 cup buttermilk

Preheat oven to 450 degrees.

Sift flour into a large bowl. If you don't have a sifter, use a colander and shake the flour into a bowl. This process creates light and fluffy biscuits.

Add the remaining dry ingredients to the bowl.

Slice frozen butter into tiny pieces. A cheese grater works great. Use a pastry cutter or fork to mix flour and butter. You want to incorporate as much butter into the flour as possible, eventually creating a crumbly texture.

Add buttermilk. Stir long enough to blend the milk with the flour. Dough will be sticky.

Sprinkle ½ cup flour onto flat surface. Pour dough onto flour. Sprinkle some of the flour on top of the dough. Press flat using hands. Use a small drinking glass to cut out the biscuits.

Bake in cast-iron skillet until golden brown, approximately 15 minutes. (continued next page)

Serving suggestion: Place a thin slice of butter in center of plate. Spoon 2 teaspoons of molasses on top of butter. Use a knife to blend butter and molasses. Serve on top of warm biscuits.

Molasses Quick Bread

This delicious sweet bread tastes wonderful on cool fall days.

Ingredients
1 ⅔ cups buttermilk
2 ½ cups whole wheat flour (or plain, all-purpose flour)
½ cup cornmeal
1 teaspoon salt
1 teaspoon baking soda
½ cup molasses

Heat oven to 325 degrees. Grease loaf pan and set aside.

In large mixer, add dry ingredients. Then add buttermilk and molasses. Mix well and pour into pan.

Bake for 45 minutes to 1 hour or until a toothpick inserted into the center comes out clean.

Run a knife along the edges of the loaf pan and allow bread to cool in pan before removing.

Baked Beans

Molasses is the secret ingredient in my baked beans. Give it a try and let me know your thoughts.

Ingredients
1 can pork and beans
¼ cup green pepper chopped
¼ cup chopped onion
2 tablespoons molasses
1 tablespoon ketchup
1 tablespoon barbeque sauce
3 slices cooked bacon (reserve for garnish)

Preheat oven to 350 degrees.

Combine all ingredients (except bacon) in baking dish. Bake 20-30 minutes or until mixture beings to bubble. Crumble bacon on top. Serve immediately.

Gingerbread Cookies

Gingerbread cookies are more than delicious; each cookie is packed with iron.

Ingredients
1 ½ cups molasses
1 cup packed brown sugar
⅔ cup cold water
⅓ cup shortening
7 cups plain, all-purpose flour
2 teaspoons baking soda
1 teaspoon salt
1 teaspoon ground ginger
1 teaspoon ground cloves
1 teaspoon ground all spice
2 teaspoons ground cinnamon
½ teaspoon vanilla

Mix the flour, salt, baking soda, and all the spices. Set aside.

In a separate large bowl, combine shortening, molasses, brown sugar, water, and vinegar. Slowly add flour and vanilla. Mixture will be stiff. If dough becomes too dry to work with, add a few drops of water. Cover and refrigerate for 2 hours.

Preheat oven to 350 degrees.

Roll dough on a floured surface. Cut into shapes with cookie cutters. Bake for 10-12 minutes. If you prefer a softer cookie, check them at 8 minutes. Cool and decorate with prepared frosting found in the grocery store. Or use your own favorite frosting recipe!

20

Harvesting and Storing the Fruits of Our Labor

When my mother was a child, her people didn't own pressure cookers. They "put up" food for the winter by drying or pickling them in crocks.

"We sealed a few things in water baths," she told me, "but we didn't own a pressure cooker until my sister, Della, got one. Most of our food was dried or pickled."

Homes of the 1940s and '50s had a crawl space, also called a root cellar, located under the house. Here they dug holes in the dirt, spread out some straw, and stored Irish and sweet potatoes. They also kept cabbages, pumpkins, and other vegetables. The space beneath my mother's childhood home boasted rows of shelves that sagged beneath the weight of blue crimp-top jars and crocks. Crockery was a common storage method in the Appalachian Mountains, as were root cellars and holes dug in the earth and covered with straw. Feral cats and outside dogs worked double-time keeping varmints out of the food supply.

"Your grandmother pickled corn and beans in a crock," my mother explained. "She used the crimp-top jars for the sausage and sometimes green beans. I remember cooking beans for hours and ladling them into jars."

Storing food for the winter was an all-day affair and required the help of every child. After a consultative view of the Farmers' Almanac to determine if the lunar signs were in the head or the heart, someone harvested the beans. Nothing was put up when the signs were in the bowels, reins, or secrets. Mercy, no. My people believed that food pickled or canned during those particular moon phases would spoil. After loading a dishpan with beans, calloused hands removed the strings, broke the beans into small pieces, and cooked them for at least half an hour. Meanwhile, the corn needed tending. Someone filled a washtub with water and stoked the fire. While the water heated, several helping hands sat in the midst of piles of corn. They removed the shucks and the silks, then placed the cobs in boiling water. Cooking time was an hour.

Remember that someone was toting all of this water from the springhouse or the creek. When my grandmother lived in the mountains—before the national park was created—there was no indoor plumbing. Winter food preparation was an outside affair with lots of heavy lifting.

After the vegetables cooled, someone cut the corn off the cob, and then it was time to fill the crocks. The process began by pouring salt into the bottom of the container. Next came a one-and-a-half-inch layer of beans, followed by an equal amount of corn. A sprinkle of salt topped off the third layer. The layering technique continued until the crock was full. Then it was time to pour in warm water. A glass plate, weighted down with a rock, sealed the crock.

Sometimes the women added peppers and cabbage. Mother calls that dish "chow-chow."

The men carried filled ten-gallon crocks under the house, where the pottery stayed until needed. In the winter, come dinnertime, someone entered the crawlspace, dipped out a serving, returned the lid, and toted a bowl of vegetables to the table.

Many homesteaders in western North Carolina had apple orchards on their property. Usually, apples ripened during the fall harvest season, leaving little time for preparation. Folk either dried or bleached the fruit. Some folk had apple presses that converted the harvest into fresh cider.

While I have never had a bleached apple, my mother swears it is the best fruit you could ever eat. She described the process: "While I peeled, your grandmother would line a box with sliced apples. She took coals from a fire and put them in a glass bowl. Then she sprinkled sulfur over the coals, put a lid on the box, and wrapped the box with a sheet. After about half an hour, she removed the lid. The sulfur bleached the apples snow white. They were delicious. We stored them in crocks and ate off them all winter. We also used dried apples for fried apple pies."

Admit it: just reading the words "fried apple pies" makes your mouth water. When I was growing up in the 1970s, my mother didn't bleach fruit, but she did dry apples. She spent hours peeling and slicing. She spread out a sheet in the middle of the yard and laid the fruit out for the entire world to see. The process took several days. Midway through the first day, honeybees located the apples. I watched them land on the slices. They carefully selected which ones they thought tasted the most delicious. Once they returned to the hive, I quickly snatched one of those slices from the sheet and popped it into my mouth. By the second day, the apple began to shrivel, compressing the juice as moisture evaporated. Trust me, modern-day fruit leather and artificial food dye-infused roll-ups taste nothing like Momma's sliced, dried apples.

❧

Made of dried beans that are still in the shell, leather britches are a traditional meal of mountain folk. Since there wasn't always enough money to purchase jars, folk strung beans on cotton thread and literally hung them from the rafters. Drying leather britches is easy. You remove any dirt, string the bean, and then run a needle and thread through the center of the bean, sort of like making popcorn garland. Do not break the beans. I've made that mistake. The first time I decided to carry on the leather britches tradition, I prepared the beans as I normally would, stringing and breaking the beans. I had no idea that they would shrink so dramatically during drying. Depending on the temperature, beans dry in a couple of weeks. Don't worry; they reconstitute when cooked like other desiccated vegetables.

When I worked for the Swain County Government, Maggie Warren was the leather britches queen. During her lunch hour, she'd string beans and then carry them to her car. She unfurled a bed sheet in the back window and displayed them in an orderly fashion. Positioned beneath the glass, the britches dried much faster than they did in the traditional string method. I fondly recall seeing her drive through town with green beans drying in the back window. She was one smart lady.

Today, few people dry beans or bleach apples. We've almost lost such traditions. I don't even know where one could buy sulfur if they wanted to rekindle this bit of history. Isn't that sad? Back then, gardens stocked the pantry. If you didn't grow it, you didn't eat. I long for the day when folk dry beans and apples in the backs of their car windows.

There is a lot of history sealed inside a clear glass jar. Few people think about the effort and emotion that goes into a jar of preserves or vegetables. There is an appreciation, an acknowledgment that we humans, despite our advances in technology, cannot force the earth to produce. We are no more powerful today than our ancestors who walked the rows and prayed that God would grant them a harvest.

Soil has absorbed tears of worry and grown hard beneath the frantic pacing of farmers who knew that their children might go hungry come winter. Work boots, slick from wear, have trekked through rows. Leathery hands have picked every single offering, regardless of how small, and tucked it beneath the house on a bed of straw. Even today, the earth collects these emotions. Some years she converts our worries into a bounty. Some years she does not. This is why Billy's statement always rings true: "A farmer is the biggest gambler around."

All of that worry, fretting, history, and hope disappear when tomatoes ripen and tiny white flowers transform into tender green beans. Then it is time for joy of ceremonial proportions.

Jars of canned goods contain prayers, pride, and so much love that the lid will barely seal. If we are honest, we admit that we have made a deal with God, or Mother Nature, saying that if granted a harvest, we'll be good stewards. We have sweated in the field while picking and in the kitchen while stringing, stirring, and baking. We have collapsed in the wee hours of the morning only to awaken and repeat the process every day until the field is bare and the pantry stocked. We have wanted to quit. We have wanted to rest. We have wanted to give half of this burdensome bounty away to complete strangers (except our springtime deal with the powers that be have made us superstitious and a tad paranoid). We can't waste a single pod, shell, kernel, or root. It's improper to give away during this phase. We must seal the gift first, twist a gold-colored ring around the jar, wrap it in a bow. Thus we continue pressing on while praying for lids to seal. We dash to the store and search for more jars, more lids, more pectin with which to preserve and complete this never-ending task our kinfolk deposited into our genetic DNA. We have had just about enough of this putting up and preserving, but, driven by guilt or perhaps self-determination, we continue.

Pride fills our hearts while we string and break beans. Later, we feel like screaming when we realize that the field contains still more. We smile as we place jars in the pressure cooker and curse when steam burns our delicate skin. We clear the counter and the kitchen table, and then, after both are full, we move the ironing board into the kitchen just like Aunt Ellen did when she was alive. Proud of our work, we marvel at the colorful jars now displayed on every flat surface. We may not remember the journey of our parents, grandparents, and great-grandparents who struggled to feed their large families, but we honor their memory the best way we know how. Surrounded by jars of every size and shape, we wipe our brows one last time, cross weary arms around our tired bodies, and give a satisfied nod before lining the pantry with the fruits of our labor.

We have made our people proud, preserved their memory, inside a glass jar.

Fried Apple Pies

Fresh apples are delicious in this recipe. You can also substitute prepared applesauce or your own dried apples.

Crust Ingredients
2 cups plain, all-purpose flour
½ cup shortening
1 teaspoon salt
½ cup ice-cold water
Vegetable oil for frying, approximately ¼ cup

Filling Ingredients
2 chopped apples (Granny Smith) or ½ cup prepared applesauce
¼ cup sugar
½ teaspoon cinnamon
Pinch of cloves

In a large bowl, cut shortening into flour. Add salt and ice water. Dough will be stiff. Roll dough into a ball and place in the refrigerator while preparing filling.

If using fresh apples, select a variety that holds up well when cooked, such as a Granny Smith. Chop apples into tiny pieces and add sugar and spices.

If using prepared applesauce, add cinnamon and cloves. Taste to determine if additional sugar is necessary.

Roll dough flat. Use a glass jar to cut a circular shape in the dough. Spoon one tablespoon of apple or applesauce into the center of the dough. Fold over and press with fork to seal.

Add oil to a cast-iron skillet, and heat until 350 degrees. Place pies in hot oil, and reduce heat to medium high. Cook 3 minutes on each side. Be careful not to burn the pies. Pierce the dough with a toothpick to determine if it is thoroughly cooked.

Remove from heat and place on paper towels.

Leather Britches

There's no way around it; in order to authenticate this recipe and cook up a good mess of leather britches, you're going to need either a hambone or a package of bean-seasoning from the grocery store.

Ingredients

A double handful of leather britches (approximately 1 ½ cups)

Large saucepan

Several quarts of water

Salt and pepper to taste

Hambone or "bean seasoning" from grocery store

Soak the beans overnight. Drain water.

Bring a large pan of water to a boil. Add leather britches and hambone. Cook beans several hours like you would a pot of pinto beans. Add extra water as necessary, but not so much that you have pot liquor. Leather britches should cook down with very little moisture remaining in the pot.

Salt and pepper to taste.

21

What to Do When Good Vegetables Go Bad

Even our favorite vegetables sometimes act out. Tomato vines become unwieldy. Potatoes and onions sprout alien-looking tendrils. This chapter explains what to do when good vegetables act out and how to increase garden yields by multiplying these unruly plants.

Sometimes onions run away. They hide in the pantry behind a can of pork and beans. You know they are in the pantry somewhere; you remember bringing them home from the store. You search until, finally, a pale tendril waves from the back, a sprout emerging from the partially desiccated, now inedible bulb. Don't toss it in the trash. The good news is that, when planted, this wayward vegetable yields "green onions."

Billy doesn't believe in the saying, "When in doubt, throw it out." His mantra is, "If it sprouts, set it out."

There are two ways to plant this bulbous vegetable in your garden: plant it whole, or divide the bulb into individual sections. Because each bulb contains individual plants that are bundled together, Farmer Billy chooses the latter.

After carefully removing the papery outside, locate the onion's core by peeling away as much of the remaining flesh as possible. The center of the onion should be pale yellow or perhaps even green. Turning the bulb over, notice any root patterns. Sometimes the roots are so entangled that it is easier to put the entire bulb in the ground and let the onion grow as it will.

The root structure determines how many "sets" each onion produces. The word "set" identifies an onion that has already begun to produce roots and sprout greenery. Wayward onions that sprout in your home contain approximately four to six sets. After locating the core, place it on a cutting board and slice it lengthwise into long sections. In order to grow, each piece requires a small amount of roots. Dig a shallow trench one inch deep and place the onion in the soil. Cover it with enough dirt to stand the plant upright. Some gardeners patiently cultivate onions from seeds. They wait a growing season for the seeds to mature into an onion of edible size.

I am not that patient, and neither is Billy.

I prefer live plants that are available from most garden supply centers, usually in bundles of fifty.

"Momma always had an onion bed," Billy recalls. "It was heavily fertilized with chicken manure and packed tight with sets. Oh, she loved her a good onion and a hot pone of cornbread."

Early spring and late fall is the perfect time to plant both the onions you've found hidden in the pantry and even the ones purchased from the store that are past eating. This technique also works for garlic. Grocery store bulbs purchased in the produce department will grow in the garden. If you cannot use an entire clove of garlic that you have purchased, don't let the cloves wither and die. For those with space restrictions, both onions and garlic will grow in a gallon-size planter. Partially fill the container with potting soil. Press the sets into the soil and cover with an inch of dirt.

A dedicated area called a "bed" provides a place where bulbs can grow without interruption. Disturbing the onion is a common mistake. When the temperature climbs to above ninety degrees, expect both garlic and onions to wither and seemingly die. Do not despair. Bulbs enter their resting phase during the summer. Keeping onions and garlic in a bed allows you to know exactly where they are located from year to year. Allow uneaten bulbs to remain in the ground. Cooler fall temperatures trigger onion and garlic to sprout with renewed vigor. Gardeners can grow several vegetables in the same area. My husband constructed planters on my deck in which I grow tomatoes in the summer and then garlic, onions, and lettuce in the fall, all in the same space. Be innovative and use every inch of dirt you have.

Potatoes are a subtle spud known for sprouting coral-type appendages almost overnight. If you have ever wondered, *Will these things grow?* The answer is, yes. The protruding parts are actually seed tubers that are seeking sunlight. "Chitting" is the official word used to define this process, although Billy prefers to say, "I've got some taters reaching for the sky. Let's get 'em in the ground."

A deep trench provides the ideal growing environment for potatoes. First, cut the spud into cubes, with each section containing a tuber. Set these aside while preparing the area. After loosening the soil and making a row, place a single piece of potato in the ground. Be careful not to dislodge the delicate shoots. Space each cube three to four inches apart and cover with soil. After leaves emerge, mound dirt around the greenery. Decomposed sawdust provides an excellent addition of organic matter, as does straw.

Potatoes grow fast. Though it depends on your zone, early spring is usually the best time to plant spuds. In the South, wait no later than mid-June. Mid-March is ideal, after the danger of frost has passed. The neat thing about potatoes is the surprise growing beneath the soil. A potato vine is short lived. For that reason, people often believe they have done something wrong when the visible plants wilt and die after a few months. But this is part of the growing cycle. Tiny blooms appear and the leaves fade. Once the vines turn yellow, it is time to grab the shovel and experience a delicious delicacy we call "new potatoes."

Some gardeners may be compressed for space, but you can plant potatoes in a five-gallon bucket. Tall containers are best. Begin by drilling a few drainage holes in the bottom. Add three to four inches of potting soil and a couple cups of sand to help with drainage. Place the cubed spuds in the soil and lightly press down so the shoots remain upright. Carefully cover with a small amount of soil. As sprouts convert into leaves, continue adding soil but no more than three inches on the top of the plant. Allow blooms to form and greenery to fade, then invert the bucket and enjoy a mess of new potatoes.

Finally, you can also grow tomatoes from ones you already have. It's actually one of the easiest vegetables to manage. Those who have grown small varieties such as cherry and salad tomatoes know that those seeds are virtually indestructible, but you can actually root pieces of the plant as well. At some point during the summer, tomatoes outgrow the confines of their cages. When this happens, snip wayward limbs and place them in a jar of water. A few days later, hair-like roots appear. Plant this new crop deep in the earth. Mulch well to extend the growing season until the first frost.

With a little creativity, it doesn't take much effort to push the growing season as long as possible. And it's always best not to let a good vegetable go to waste!

Onion Soup

By far, this is the simplest soup ever, with a rich broth and hearty flavor. If you like onions, you will love this recipe.

Ingredients
1 large onion, sliced thin
2 cans beef broth
½ cup water
¼ teaspoon pepper
1tablespoon olive oil

Drizzle olive oil in a cast-iron skillet, then turn heat to high. Once skillet is warm, add onions. Cook for 1 minute on high, then reduce to medium and cook 3 to 4 minutes or until onions are translucent.

While onions are cooking, pour broth into pot and turn on high.

Add translucent onions to the broth, then use ½ cup of water to deglaze the skillet and capture any remaining onion juice. Pour into pot of broth. Add pepper. Reduce heat and serve with garlic chips (see recipe below).

Garlic Chips

Turn plain flour tortillas into delicious chips in a matter of minutes.

Ingredients
1 bag flour tortillas
3-4 tablespoons mayonnaise
3 tablespoons chopped garlic

Heat oven to 350 degrees.

Cut tortillas into triangles. Lightly spread mayo on tortillas then spread chopped garlic on top of the mayo. Bake for approximately 5 minutes or until triangles are lightly brown. Serve immediately.

Onion Dip

Come game day, there is no need to purchase powdery onion soup mixes. This simple dip takes about three minutes to prepare, doesn't contain MSG, and tastes delicious.

Ingredients
1 large onion, finely chopped
1 small container sour cream
Salt and pepper to taste
1 tablespoon olive oil

Add olive oil to skillet. Place onions in skillet and cook until they turn golden brown. Allow to cool.

Pour container of sour cream into a bowl. Add a pinch of salt and pepper. Stir in onions. Serve with chips.

Wasn't that easy?

Grilled Onion

Here is an easy way to enjoy those famous Vidalia onions. You can grill them, or bake them in the toaster oven.

Ingredients
1 large onion
1 pat butter
Pinch of salt and pepper

Peel onion and cut into quarters. Place onion on a sheet of aluminum foil and dot with butter. Sprinkle with salt and pepper. Twist foil around onion and place on grill or in a 350-degree oven. Cook for 5 to 10 minutes. Test doneness by squeezing foil.

Unwrap and serve immediately.

New Potatoes

Fresh and tender, new potatoes are always delicious.

Ingredients
2 cups freshly dug potatoes
2 quarts water
Salt and pepper to your taste
2-3 tablespoons cooking oil
1 teaspoon garlic powder (if desired)

Bring two quarts of water to boil. Add approximately 1 teaspoon of salt to water. While waiting for water to boil, scrub potatoes to remove traces of dirt. Leave skin on. Cut potatoes into small cubes. Cook for five minutes or until you can pierce the skin with a fork.

Drain. Heat cast-iron skillet and add enough cooking oil to cover the bottom of the pan.

Add drained potatoes and reduce heat to medium. Cook for 5 minutes or until skins are lightly brown. Season with salt, pepper, and garlic powder if desired.

Ball

22

My Parents Visit the ATL

Cancer brought my mother and Billy together. They became pen pals after both experienced cancer scares. They exchanged get-well cards and inspirational scriptures, and they uttered prayers whispered between interwoven fingers. Billy has won the battle with prostate cancer. Mother continues her war with ovarian cancer.

My parents don't get out much. Their home hosts a flock of chickens, a dog, a few cats, and a grandson who visits each day. During my ten years of living in Atlanta, they have visited once. So when I received the call that they were coming down for a visit, concern replaced my initial joy.

Something must be wrong. Folk from Appalachia tend to stay at home. They don't drive to the big, big city unless something is terribly, terribly wrong.

"I think it's time for me to meet Billy," Mother said.

Mother is a private person. She commanded that I not share her personal business with the world, especially not on Facebook. I imagined her trip was because someone in the family had news of the depressing variety.

"We'll come down, spend a little time with Billy, spend the night, and then leave the next morning."

Overnight visits are highly suspect. Mountain folk are uncomfortable away from the protection of hills and hollers. Who would feed the chickens and gather the eggs?

I had previously programmed Dad's GPS with my address, just in case he got turned around in the metropolis of Asheville during one of Mother's doctor visits. I saved my address under ADAUGHTER. No search needed, just press the first address stored alphabetically in the memory. Mother, on the other hand, doesn't trust newfangled technology. She insisted on a printed map and turn-by-turn directions, complete with the names of the gas stations that have the cleanest bathrooms.

"Now, don't plug in the GPS until you get to the Walmart in Murphy," I warned, referencing a familiar landmark, "or you'll get all mixed up.

Having opted to route them as far away from the interstate as humanly possible, I knew that the female voice inside the GPS gadget would insist they turn around if they powered her up in their gravel driveway.

My parents arrived in one piece. I took them to the farm, where Billy and Mother embraced like family and brushed away a few tears. They were kindred spirits bonded by prayer. Billy broke out all the family albums that his daughter had assembled over the years. This triggered stories about the good old days.

Like most visits, there were things to do on the farm. Dad coveted the tractor while Mother enjoyed a garden tour. As expected, she fell in love with the goats.

Billy opened the hay compartment located beside the chicken house. Inside, he scraped the contents of a brown wagon into a cracked trash can.

"The weatherman's calling for rain," Billy said, then added, "Tomorrow I'll need to get some hay before it comes."

Surveying the field, Mother asked, "Where do you buy hay around here?"

Billy and Mother chatted and fed the goats while Dad and I moseyed through the property. "You know, I'd like to work on the Chevy," Dad whispered. "I believe I could fix it."

I agreed. Dad can fix anything. It is his gift and his curse. Family and friends rely on his handiness, to their benefit. He really should charge, at minimum, for the parts. He and I believed he could transform Good Old Reliable to her former glory. The only thing Dad needed was the opportunity and perhaps more chewing tobacco.

That night at dinner, Mother said, "Billy told me how he gets the hay. Do you know he scrapes up the loose pieces from the back of a tractor trailer?"

I nodded as Mother said, "Larry, I think you should help him. We'll get up early tomorrow; you and Renea can help with the hay. We'll leave after y'all are done."

She often volunteers Dad's services. If we had a nickel for every time she has said, "Larry can fix it," we'd be sipping an umbrella beverage and spitting tobacco juice.

The beverage is mine; tobacco juice is Dad's.

The following morning, we returned to the farm. Billy was polishing off the last bite of a sausage biscuit. Dad smiled, pleased that Billy enjoyed his breakfast. The meat was a gift, preservative free. Wild hogs grow fat on grain and acorns in the mountains where my parents live. As long as the swine knew their place, Dad was fine to let them roam. But a wild boar has a propensity toward pushiness. As Dad's corn-growing season progressed, the pig moved closer toward the corn. Once the swine cut a swath through the field, it was all over but the aiming.

Feral pigs are dangerous nuisances. They can quickly destroy property, uprooting grass and shrubbery. Their snouts create divots and then trails that encourage soil erosion. One might think hogs call the deep woods home, but not these days. With man as their only predator, the population has grown exponentially and now expands into residential areas. Let's be honest: how many people do you know with the nerve to drop a wild pig?

K & B Meat Processing, a company in Bryson City, North Carolina, processed the pork soon after Dad's gunshot. K & B have been in business for as long as I can remember, locally owned by wonderful people. The meat contains no preservatives and is healthier than the store-bought sausage Billy purchases in the reduced section of the grocery store.

ঌ

Mother stayed behind and stoked the fire while Dad, Billy, and I strapped ourselves into the new F150 and headed toward Ball Ground, Georgia. Our final destination: Cherokee Feed and Seed, where we would collect hay for Billy's critters. We were barely out of his driveway, the house still fading in the rearview mirror, when I observed the roadside scenery passing by at a rather brisk pace.

"Uh, Billy," I said cautiously, "do you know the speed limit?"

I asked this question at the inopportune moment when the traffic light changed from green to yellow. He answered with a frantic stomp on the brake.

Whiplash. That's what I feared as a water bottle crashed to the floor and my body rushed forward and then slammed back into the seat. *Soon I would have a healthy case of whiplash.*

"Good night! Looks like you'll need to hold me back if we're going to get there without a ticket," Billy replied.

My talkative father fell unusually quiet.

The possibility of a ticket triggered a story about the last time Billy visited Cherokee County. "I was driving along this here road like I always do, when I noticed a blue light flashing behind me."

Sensing the beginning of a good story, Dad twisted the cap off an empty plastic bottle and spat tobacco juice inside.

"I saw the blue lights and I reckoned the police officer had a call. So I pulled over to let him pass. Would you believe he was after me?"

Billy took his eyes off the road and looked over to Dad. I tightened my grip on the seatbelt and prayed that someone other than me was watching where we were going.

"So I rolled my window down to see what the police wanted. The officer said, 'Buddy, you're doing 59 in a 45 zone.' I told the officer, 'I was just driving like I always do.'"

Billy paused. A smile formed on his face. "By the way, I got off with a warning."

After checking the side mirrors, he looked at me through the rearview and said, "You'll need to keep a watchful eye, at least while we're in Cherokee County. They're after every dollar they can get."

From the back seat, I really couldn't see as well as Dad, but I didn't argue.

"Yeah, but you're in this fancy truck," I replied. "You drove Good Old Reliable last time, and the officer was probably shocked it could go that fast. This truck will get you a ticket."

I noticed that the speedometer held steady at straight-up sixty miles per hour. I uttered another gentle reminder.

Billy exclaimed, "Good night. Sixty!" He transferred his foot to the brake pedal and calmly said, "Oh, well. If we make it there without getting caught, we'll surely get there a whole lot quicker. Maybe I'll be alright if I just go five miles over."

While driving through what he called the "Birmingham Community," Billy pointed to an unassuming, unoccupied block home and said, "Me and Marjorie lived there from 1940 to 1945. Would you believe we paid twenty dollars each month in rent?" Adding an incredulous, "That was two hundred and forty dollars a year." Billy retold how he and Marjorie scrimped and saved until they had enough money to build their own home.

"Now the house wasn't finished when we moved in," he clarified, "but it was a roof over our head and it was paid for."

As we traveled past the gas station, he admitted he had a "hard time understanding the world today." Pointing to the price of gas, he said, "Look at that, gas is $4.00 a gallon. Can you believe that?"

Truly, the price is an outrage, regardless of your age.

Dad said, "Twenty years ago, if someone had told me that people would buy water in a plastic bottle, the same water I got for free out of a creek, I would have called him a crazy man."

Billy nodded. "By the way," he said, "I'm going 60." Another tap on the brake. "This thing shore is pert."

Arriving at our destination in record time, Billy pulled into the gravel parking lot and parked near the back of the building. Dad and I followed. Once inside the storage room, I allowed the men to go on ahead. Frozen for a moment by the smell of oats and animal food, I stood as the aroma pricked my heart and transported me to my grandpa's feed room. Closing my eyes, I inhaled the remembrance, still heartbroken that both Grandpa and his country store named Winchester's Grocery are forever gone. Wiping my eyes, I opened the door and joined the men. Awestruck at both the selection and the cleanliness of Cherokee Feed and Seed, I quickly identified the owner as a three-pound canine sitting behind the register. While her minions rang up the sale and worked around her, she observed. Employees of Cherokee Feed and Seed care about animals, the community, Billy Albertson, and others like him.

I much prefer this place to somewhere like Tractor Supply Company. Even though that company is American-owned and boasts that their customers can find everything they need, I find their shopping environment sterile, cold, and impersonal. Farm supply stores need separate feed rooms, with chairs arranged in a semicircle. They need ashtrays or spittoons in the middle just like the good ole days. Friday night pickin' and grinnin', turkey shoots in the fall, baby chicks come spring. Absent that, Tractor Supply should at least smell like a country store.

Cherokee Feed and Seed's storage room is efficiently stacked, floor to rafters, with every type of kibble imaginable. Genuinely courteous, the employees offer fantastic customer service, and their pride shows in their selection and the friendly way they do business. Just try to load your own animal food. I dare you. Reach down and act like you're going to grab a bag. Trust me. You will see chivalry at its finest. You will hear, "I got this, ma'am. This is what they pay me for."

Being called "ma'am" makes me uneasy—I haven't yet reached the age of maturity—but, when it comes to toting fifty-pound bags, color me ma'am. Odds are, at your next visit, workers will remember your name, perhaps even recall the brand you purchased. You won't find that at stores like Tractor Supply. I promise.

Once, and I do mean *only* once, on my way to North Carolina, Dad asked me to pick up a fifty-pound bag of scratch feed at the Tractor Supply. Familiar with feed stores such as Cherokee Feed and Clampitt's Hardware in my hometown, I paid for the feed and then waited for the clerk to summon an employee from the back to tote the bag to my car.

Peering around the front of the store, I said, "I'll go pop the trunk while you call someone to load my purchase."

The cashier stared at me like I had three heads. Standing on tiptoes, she pointed to the back of the store and said, "Feed's in the far back of the building. Buggy's over there."

You gotta be kidding me. Who in the world had the brilliant idea to stock the heaviest merchandise in the back of the store?

"Should I drive around back?" I asked. "Wouldn't it be easier to load from there?"

Shaking her head, she said, "Can't. No back entrance."

Right then and there I swore off Tractor Supply. Trust me, the fifty cents I saved was not worth the back strain. Heck, I would tip the guys at the local store at least a dollar, but the canine queen won't allow it.

ꝏ

We were there for hay. Not the baled variety; we wanted the dregs. But first, Billy paid for two bags of goat feed and got into a brief discussion with an employee who insisted, "Now Billy, it's my job to load your vehicle. You just head off toward the hay before you git me in trouble."

Waving his hand, Billy said, "Aw, go on then. Load her up."

Even Billy is afraid of the three-pound canine owner.

At Cherokee Feed and Seed, tractor trailers deliver hay. As employees load the bales, tiny pieces dislodge and accumulate on the trailer floor, creating a slippery, hazardous surface. Someone must remove these bits of hay before refilling the trailer with new bales. Allowing Billy and others to clean the floors benefits everyone, especially the animals.

Billy cranked Tinker and situated the truck at the mouth of an open trailer. It takes several helpers to collect loose pieces of hay. Usually, Kelle and her three boys have this privilege, but she was out of town, and, according to Billy, he needed to "make hay while the sun is shining." Meaning rake now before the gathering storm clouds dropped buckets of rain.

Billy tossed two claw-shaped rakes with tines the width of a man's hand out of his spotless truck. Handing two leaf rakes and the snow shovel to Dad, he instructed, "Prop the step ladder against the trailer."

Climbing into the trailers requires cat-like agility. While he climbed into the trailer bed, Billy instructed me to remain inside the truck. My job: drive to the next trailer when told. Working from the back of the container toward the front, he eventually emerged with his arms full of loose hay, which he dumped into Dad's outstretched arms. The process was as slow as Christmas. Abandoning the F150, I took one of the rakes and climbed into the trailer bed with them the men. Within moments, my throat burned and my skin itched. I struggled to breathe. I am terribly allergic to grass. Realizing this, instead of staying home with Mother, I took medication. She had offered a mask she wears when she receives chemo. At first I said no, but she insisted. As I adjusted the mask over my nose and mouth, I decided I was grateful to have it. Billy should wear one too. Before long, he was also sneezing and coughing. Alas, he declined my offer.

Metal tines bounced across a wooden floor. Plumes of dust billowed and hovered inside the enclosed area. Without means of escape, the particles danced in the air. After determining a faster way, I abandoned the trailer and worked one ahead of them. Climbing in, I pressed a snow shovel firmly to the wooden floor and then walked quickly toward the open doors. As I kept the shovel pressed against the wood, dry grass rolled into the shape of a ball and accumulated at the edge. After three efficient swarps of the blade, I was done. Staying two steps ahead of Billy and Dad streamlined the gathering process and significantly decreased the amount of throat-choking dust we inhaled. The men needed only to drive the Ford forward, toss hay into the bed, and then continue to the next mound.

When we had finished, the interior of Billy's pristine new ride was tarnished. Tiny particles of hay had fallen from our clothes and imbedded in the carpet. Despite our attempts to kick clods from our shoes and shake debris from our clothes, streaks of mud speckled the floorboards. Tinker was officially a work truck, baptized with Georgia clay and alfalfa.

Ideally, purchasing intact hay bales is the most effective use of my time. In today's busy world, the four hours it took to drive, rake, and pack the truck were excruciating—especially when factoring in the time it took to unload the hay after Billy returned to Hardscrabble Road. This investment of time does not equal free hay. However, in this hurried world in which we live, I want to learn to find enjoyment in the slower things. Besides, this chore afforded Billy the opportunity to do a little tradin', a little politickin', and a lot of driving fifteen miles over the speed limit.

By the way, Dad drove the Ford back home.

Coconut Cake

If you have never made a cake from scratch before, try this recipe. This is my dad's favorite.

Cake Ingredients
1 cup sugar
½ cup shortening
3 eggs, separated (reserve whites for icing)
1 teaspoon real vanilla
2 cups plain flour
½ teaspoon salt
1 teaspoon baking powder
½ teaspoon baking soda
1 cup milk

Icing Ingredients
1 ½ cup sugar
3 egg whites
⅓ cup cold water
1 tablespoon Karo syrup
Pinch of salt
2 cups grated coconut for garnish

Preheat oven to 350 degrees. Grease 2 (9-inch) round pans and lightly sprinkle 1 tablespoon of flour in the pan to prevent sticking.

Cream the sugar and shortening. Add egg yolks. Mix well and scrape sides of bowl. Stir in vanilla. Add flour, salt, baking powder, and baking soda. Mix well. Add milk, and mix well.

Bake cake for 40 minutes.

(continued)

Make icing using a double boiler. Place sugar, egg white, water, and syrup in a bowl. Using a hand mixer, beat until light and fluffy, approximately 7 minutes. Add a pinch of salt. Frost cake after it has cooled.

Add coconut between layers, and sprinkle on top of cake and sides as well.

Glazed Pork

This quick and easy dish is delicious.

Ingredients
1½ pounds pork tenderloin
3 cloves garlic
2 tablespoons light brown sugar
2 teaspoons steak seasoning
¼ cup red wine
2 tablespoons butter

Crush garlic and place in small bowl. Add brown sugar and steak seasoning to bowl. Mix well. Cut pork diagonally into 1-inch-thick slices. Place pork in bowl of seasoning, and coat both sides with spices.

Preheat large pan on medium heat.

Place one tablespoon of butter in the pan. Add pork. Cook 4-5 minutes on each side or until the temperature reaches 160 degrees. Use a meat thermometer to determine temperature.

Remove pork from pan. Deglaze the pan using the remaining 1 tablespoon of butter and the wine. Stir for one minute. Return pork to sauce. Serve immediately.

Liver Mush

Before vegan was vogue, liver mush was a Southern staple. For many years, Billy worked as a butcher. Souse meat and liver mush were two of their big-selling items. Today, finding the key ingredient to make this recipe is as difficult as locating hen's teeth.

Billy likes to eat his liver mush cold. He carves off a hunk and wraps a piece of white bread around the meat.

Shudder.

My friend Rachel grew up eating Hunter's liver mush, which is proudly produced at 98 Poteat Road in Marion, North Carolina. The Hunter family has been satisfying their customers since 1955, and Hunter's liver mush has fed hardworking folk for generations. The Hunter girls took piano lessons from Rachel's mother and sometimes paid for their lessons with hunks of liver mush. Rachel prefers slices that are fried crisp and adorned with a hefty dollop of ketchup.

Did you know that several towns across the South have liver mush festivals? Now that's heritage, my friends.

Ingredients

2 pounds pork liver
1 cup cornmeal
4 cups water
1 tsp salt
½ teaspoon pepper
1 teaspoon sage
¼ teaspoon cayenne pepper
1 egg
1 tablespoon olive oil
Small loaf pans

Rinse liver, then sprinkle with salt and pepper. Pour olive oil into cast-iron skillet. Sauté liver over medium heat, cooking until done (until juices run clear). (continued next page)

Allow to cool. Place in food processor and finely chop.

In large saucepan, bring water to a boil. Add cornmeal, sage, and cayenne pepper. Mix egg into liver, then stir into cornmeal mixture and cook for five minutes. Taste and add extra salt, pepper, and sage.

Spoon into small loaf pans and bake at 350 degrees for 30 minutes or until set.

Moon Pies, A Family Tradition

People who aren't from the South may not understand the importance of this much-loved snack cake. In the 1930s, a working man could purchase a ten-ounce RC Cola and a Moon Pie for a nickel apiece. The double-decker Moon Pie became available when I was three years old. It boasts three cookies and two layers of marshmallow. Glory be, nothing on earth is better than enjoying one of these with your people. Every bite is a delicious tradition.

Ingredients
One Moon Pie (the Winchester family prefers banana flavored)
One microwave
One minute of your time, less if your microwave cooks fast

Unwrap the pie and place it on a plate.
Press start.
Watch the pie. When the marshmallow puffs, remove from microwave and enjoy.

23

Heritage Corn

For some reason, corn brings out the worst in Billy's customers. Otherwise friendly folk go rogue. They become rude, impatient, and downright un-Southern—all for a dozen ears of Silver King.

After noticing a stack of crates piled beside the plunder building, I asked Billy, "What are you doing with these?"

"Oh, one of my customers wants me to save him seventy-two ears of corn," Billy replied.

"Seventy-two ears! You can't let one person have that much. There won't be enough for the rest of your customers."

C'mon now, folks, let's not be greedy. We only have a limited amount of corn.

During a particularly hot summer, when the temperatures peaked at 106 degrees, customers were so rude that I threatened to cut down every single stalk. It was a code orange day. The air contained so much unhealthy particulate matter that the morning news warned people, especially the elderly, to stay indoors. I had the feeling I should check on Billy, who doesn't have air-conditioning. Moments after he answered the phone, I almost had a full-blown conniption when he said, "I'm out here pulling corn for a couple customers who just showed up."

A conniption occurs when anger bubbles up from within and shoots out the top of your beehive hairdo. Think Mount St. Helens, or when someone asks for change from the cashier at the toll lane.

Outraged, I screamed into the phone, "It's one hundred and six degrees! I can hear you struggling to breathe. Didn't you see the news? You are not even supposed to be outside, much less picking corn."

Leaving him no time to respond, I added, "I am getting in my car right now. If I find you in the garden when I get there, there's gonna be trouble."

Billy's voice muffled as he turned to the customers and said, "Folk, it's too hot out here. I think we need to get inside."

You're darn tootin' he needs to get inside. What his customers needed was a good ole-fashioned talkin' to.

As Billy once told me, "A good talkin' to did me a sight more good than a whoopin'."

Dialing Janet's number, I tattled, "It's hotter than Hades, and your father is outside picking corn for some customer. Could you please call the man and make him go inside before I go over there and cut down all the corn?"

Allow me just a moment here. Who in their right mind pulls into Billy Albertson's driveway at three o'clock in the afternoon when the temperature is dangerous and then demands corn?

ꝏ

My rural upbringing leaves me yearning to plant heritage seeds. I also have a fierce desire to stay true to my roots despite living in a metropolitan area. My deck planter grows tomatoes, beans, and anything else I get my hands on. The deck isn't glamorous, but I can look outside on any given day and remember people who blistered their hands just so I could write their story. If I don't honor the memory of my people, their stories might be forever lost.

And I will have none of that.

My dad was on a mission to find a particular type of heirloom corn he called "white prolific." He wanted this variety because his grandfather, Columbus "Lum" Winchester, had planted it when my dad was a kid. Lum Winchester had a reputation of being the best corn grower and blockader in the area. A blockader is old-timey speak for someone who makes and sells moonshine. My great-grandpa was clear when he talked to the news boys about his younger days; he was a blockader.

A story about Lum Winchester written by John Parris appeared in *The Asheville Citizen* in 1979. My sweet great-grandfather, who was ninety years old at the time, explained that he had "took up [blockading] when I come back from the war across the waters. About the only means a feller had for payin' his taxes was whiskey makin'."

A lot of folk believe blockaders were toothless, tobacco-juice-spittin' yokels. My great-grandfather was a smart businessman. Once he got out of the likker-making business, he made money raising cattle. He didn't drink up his likker profit. Instead, he invested it in land because he knew that one day land would be hard to come by. Even though he was the father of nine children, money wasn't all that important, but family, now, that was everything.

Searching the World Wide Web, I uncovered what I believed were seeds that would make my dad proud. We didn't plan on pressing and distilling corn. We just wanted to grow a little bit of history. Clemson University has an heirloom seed collection consisting of several non-hybrid varieties. These seeds come from Dr. David Bradshaw (Retired) of the Horticulture Department. At first I wasn't certain I had located the corn I wanted. It was named "John Haulk Corn." Mr. Haulk was a farmer who had grown the corn for over fifty years. Scanning the information, I learned that Dr. David Bradshaw obtained this corn from Oliver Ridley.

Inside my brain, the bells chimed.

Ding. Ding. Ding.

My great-great-grandmother, Mary Magdeline Ridley Winchester, was Lum Winchester's mother. This made perfect sense. Where else would he receive the seeds? I believed that I had found the same corn he planted. I was so happy I could cry. Dialing my dad, I learned that he had also been successful. While visiting the farm of Roy and Rachel Griggs, Dad had made a comment about the beauty of their corn. Roy and Rachel are real homesteaders and are undoubtedly some of the best folk on this planet. I have known the Griggs family since God put me on the earth. When I lived in North Carolina, we attended the same church. They live next door to my people. So I guess Dad shouldn't have been surprised when Roy said, "That's Lum's corn."

Moments later, my dad received a grand gift, an enormous ear of corn.

"I like this corn," Dad said, "because you never know what the ears will look like. Sometimes the kernels are solid white, sometimes pink, sometimes a combination."

John Haulk corn is open pollinated, as is Lum Winchester corn. This means the corn grows in fields where it either self-pollinates or cross-pollinates with other corn. Open-pollinated seeds are desirable for the home gardener who wants to save seeds from year to year. Because of the space required, this particular variety yields fewer ears than hybrid. Most hybrid corn is genetically altered to produce multiple ears on a single stalk. It is also designed to grow close together. Not so for John Haulk corn. It needs plenty of space. The stalks grow at least ten, sometimes fifteen feet tall, dwarfing hybrid varieties. When given adequate room, each stalk yields two ears that are twelve to fifteen inches long. Harvest requires either a ladder or a machete. Dried kernels are flavorful and make delicious grits and cornmeal. While many farmers grow this corn for animal feed because of its nutritional value, once restaurants in the Low Country discovered the rich flavor each kernel contained, chefs demanded John Haulk corn.

Proudly displaying a tiny bag of kernels, I told Billy the history of the corn and then begged him for enough space to plant just a couple rows.

I think my great-grandpa is proud.

ꝏ

Neither Billy nor I knew the name of the men who visited during summer 2012. Billy welcomed them and gestured toward the produce stand, where they were invited to have a look around if they were so inclined.

Boy, were they ever.

Instead of selecting produce from the table, these two men walked straight to the garden and ripped the ears from my Lum Winchester cornstalks.

Lawd, have mercy. Somebody is gonna get it.

Making their way toward the front of the property, the men paused where I was stringing tomato vines and proudly displayed their arms full of corn.

My usually jovial personality exploded. "Did Billy say you could pick that corn?" I demanded, knowing full well that no one had permission to pick the corn. It was months away from being ready. Field corn is the tortoise, and sweet corn's the hare. Field corn does not taste like sweet corn that is served with butter, salt, and pepper. It stays in the field until it gets good and ready, sometimes taking up to one hundred days or more from germination to harvest.

"He said we could make ourselves at home," one of the men said.

"That corn is feed corn," I explained. "It is not ready to eat."

Without apologizing, the other brother spoke to me as if I were ignorant about the ins and outs of growing instead of someone who has blistered her hands working the soil.

"It certainly is ripe," he said while shaking an ear in my face.

This was the exact moment I smelled alcohol on his breath. Did I mention it was 10:30 in the morning? I found it ironic that an intoxicated man harvested my great-grandfather's likker corn. The inebriated man shoved an ear of corn in my face, ripped the green leaves back, and, even though the cob was devoid of kernels, said, "This is ready to eat."

He was obviously not from around here.

"Don't mind him," the other brother defended. "He thinks he has to be right all the time."

Not the least bit intimidated, I pointed my finger at the man who towered above my five-foot-tall frame and said, "That is my corn. Not Billy's. I planted it. It will not be ready until October. You should not have picked it."

"But it's ripe," he defended, his breath sour and strong.

Stepping closer, I said, "I'd better never see you in the corn field again. Do you understand?"

I think he did. At least the sober man understood. Taking his intoxicated brother by the arm, he led him away from me.

A few minutes later, feeling bad for my outburst, for pitching a fit about the corn I had grown on Billy's property, I asked Janet if she knew the men.

She responded, "No. They left without speaking to me."

"How much did they pay for the corn?" I asked.

Puzzled, Janet said, "They didn't pay me. I thought they paid you."

Lawd, have mercy. Now somebody is really gonna get it.

"You're kidding me," I said as my anger bubbled. "Here's the deal. First off, one of them had been drinking. He reeked. I could smell liquor on his breath. Then they raided the field corn, took stuff that isn't even ripe, and now you're telling me they didn't pay?"

Placing her hands on her hips, Janet replied, "Apparently not. Who were those men?"

"I don't know," I answered. "I hoped you knew them."

Billy didn't know either. When I told him what had transpired, he said, "Now you know I can't control how people act on my property."

That may be true, but I can ask those who misbehave to leave.

Grabbing a permanent marker, I constructed a sign: LUM WINCHESTER CORN. NOT FOR SALE. PLEASE DO NOT ASK.

"Please don't sell my corn," I begged Billy. "I want to save the seeds and grind it into meal. It's my great-grandpa's heritage. It's very important."

I know he understood.

Growing heritage corn takes patience, a sunny spot, and not much maintenance. From the moment the nodal roots take hold until the last tassel turns brown, the Lum Winchester crop grows tall, shading the garden floor. First-timers wishing to cultivate a hill or ten should understand the simple growing process. Once delicate plants reach three inches high, it's time to "work out" the rows. Those with tractors drive machinery into the field, dislodging any weeds. Even so, someone must still grab a hoe and hack down the invaders who cling close to the stalks, sucking precious nutrients from the soil. Commercial farmers sometimes spray chemicals during this phase.

Shudder.

Using a hoe, farmers uproot weeds and then pull dirt around each plant, creating a mound. This process, commonly called "hilling up," assists brace roots that appear later during the growth cycle. During this time, farmhands apply a liberal amount of fertilizer. Dad prefers 17-17-17. Billy uses 10-10-10. Applying a handful beside each stalk, one understands what it means to "side-dress the corn." Here's a helpful garden tip: before a rain is the best time to work the corn out and apply a quick side dressing. Moisture carries nutrients to hungry roots. Soon, your corn will be like Billy's, strutting along peacock proud.

While Billy enjoys the crop because it brings in customers, folk like Noemi have an emotional attachment to corn. I met Noemi, the Tamale Lady, at Big Shanty Antiques & Gifts in Kennesaw, Georgia. Tucked inside a charming store, Noemi works twelve-hour days wrapping her soul inside moist corn shucks.

I get excited about meeting new people and learning their success stories. When a friend told me about Noemi, I was thrilled at our chance to meet. Some people exude love. Noemi is one of those people. After instructing me to wear a hairnet, she patted a stool.

"Now, I can't give away my secrets," she began as I settled in for a lesson.

Nor would I ask her to do so.

She talked as she worked. Her gloved hands grabbed a shuck painted with masa, which is a corn-based dough. Then she added the meat, cheese, and her secret ingredient, love. I could feel it. Pure love, almost an anointing, pressed down tight and then folded, wrapped, and steamed inside each tamale.

"I don't want to offend you," I confessed. "My husband likes tamales, not me. I've had them before, and they're just not my thing."

She nodded. "I've heard this before. But people have also said, 'This is the best tamale I've had in thirty years.'"

"Why make tamales?" I asked. "Why make something that is so labor intensive and takes so long to prepare?"

Without pause, she said, "Because this is who I am. I've been making tamales since I was eight years old. My mother was raised in Aspermont, Texas. Every Christmas she made Boston butt tamales. I would get them out of the steamer and put them on plastic plates, then serve my brothers. When I lived in San Antonio I continued the tradition. I love feeding people."

"I am glad I grew up poor," she said, crumbling cheese on top of a skirt-shaped husk. "I load my tamales with cheese. We couldn't afford it when I was growing up. Being poor was good for me. I learned how to work hard."

Those who have made tamales understand the definition of hard work. Personally, I would never attempt this process. Building these flavorful bundles literally takes all day.

After quickly folding a husk, Noemi reached for another, "This is who I am and why I make everything from scratch." Turning to me, she asked, "Have you eaten lunch?"

When I confessed that I hadn't, she placed a plate before me. She carefully removed the foil and opened her gift. Steam rose. My mouth watered.

She loaded a fork and said, "Here. Taste."

Spices blended with a refreshing splash of lime. The meat was tender. Moist. Masa was perfect, not thick or soggy like others I had tried.

"Yum," I said.

Noemi smiled and said, "Exactly."

She understands tradition, like why I grow my great-grandfather's corn. Just like I understand why she must make her mother's tamales and offer them to you.

&

Corn is a waste-not crop. Billy uses every single shuck and kernel to feed either man or beast. After removing the developed ears, he leaves the rest of the plant to dry in the field until he announces, "It's fodder-pulling time."

Fodder is old-timey talk. It's what farmers fed their animals: dried leaves of corn removed from the stalk, wrapped, and hung in the barn for later use.

"We used to feed the mules five ears of corn and a shock of fodder every day," Billy said.

Most commercial farmers feed their animals silage during the winter. Think of silage as one big silo full of pickled corn with a little sugar mixed in for good measure. Stored green and preserved through pickling, this chow is fermented by natural sugars and will keep up to three years. Perhaps you've seen machinery chop everything in a field, spitting pieces into an open truck that travels alongside the tractor. That's silage in the making: cobs, leaves, stems, grain, and most likely sugar cane are tumbled together. Harvesting after the cobs fully develop, and before frost, the machine grinds everything down to bite-sized pieces. Cattle think that silage is the best thing since sliced bread.

Sharecroppers like Billy's Poppa rarely rented silos or had access to heavy equipment. They dried animal food in the winter and stored it in barns.

Following behind Billy, I watched him grab a leaf at the base of the stalk, give it a hard yank, and then continue down the row until one hand was full. Using the longest leaf, he twisted and tied the bundle and then hung it on the naked stalk.

Copying his method, I watched as the entire stalk bent forward, leaf intact.

"Here, you gotta pull straight down," Billy told me, "like this."

With one hand full, I looped the leaf, tied, and failed. Bending over, I wedged the bundle between my legs. Then, using both hands, I wrapped a leaf around everything, looped, and tried again only to shred another leaf. "Billy, I can't figure out how to tie this. I keep breaking everything."

After all these years, I still can't do half the things he does.

"Your shock is too thick. Just get one handful. Here, see?"

Like a magician, he twisted and tied and then presented the shock. The only thing missing was a "voilà."

"How about I pull and you tie?" I finally suggested.

Billy agreed. I didn't understand how the desiccated leaves possibly contained any nutritional value. But, working together, we pulled the leaves, wrapped them, and then hung them in the shed.

"Now, you make sure you write about fodder," Billy instructed. "Take a picture too. Folk need to know that back then, we never wasted anything."

He still doesn't waste anything.

Corn and Bean Medley

This is a delicious side dish that you can serve any time of the year.

Ingredients
1 can kernel corn, drained
1 can black beans, rinsed
1 small onion, chopped
1 bell pepper, chopped
2 sprigs fresh cilantro, chopped
2 tablespoons ranch dressing
½ teaspoon cumin
Salt and pepper to taste

In large bowl, add corn, beans, chopped onion, and chopped bell pepper. Toss well. Add cilantro, ranch dressing, and cumin. Add salt and pepper if desired. Serve as a side dish or with corn chips.

Cream-style Corn

Some folk also call this "creamed corn." Regardless, this delicious dish leaves you wanting more and longing for home. Pay close attention to the proper cutting technique listed in the instructions. Incorporating this technique releases more sweetness from each kernel.

Ingredients
4-6 ears fresh corn
2 tablespoons real butter
1 teaspoon cornstarch
4 tablespoons water
1 teaspoon salt
Black pepper to taste, if desired

Remove husks and silks from the corn.

Cutting the corn takes three passes. Hold the cob over a bowl. With the first pass, remove just the top of the kernel, and with the second slice remove the middle portion of the kernel. For the third cut, place the knife vertically and hold it as close to the cob as possible. Then scrape to remove the sweet juice.

Place the corn in a saucepan, and add just enough water to keep the mixture from sticking to the bottom of the pan (up to 4 tablespoons). Add butter and salt. Cook on medium heat. Stir often. If corn sticks to bottom, it will burn and ruin the entire dish.

Cook for 10 minutes, then taste. Add more salt if necessary and pepper if desired. If corn is too thin, thicken with cornstarch. Place 1 teaspoon of cornstarch and 1 teaspoon of water in a small bowl, mix well, and then pour into corn. Stir well.

24

Puerto Rico

Human nature predestines us to judge people based on their appearances. We put people in categories based on their clothes, skin color, hair color, size, and facial expressions. Many visitors look at Billy's overalls and frayed straw hat and see a simple man. Billy Albertson *is* a simple man. He resoles his shoes with duct tape and staples. He recycles plastic sacks and cardboard egg cartons. He doesn't require worldly possessions. But, thanks to his daughters, he has also done quite a bit of traveling. Sometimes he visits the world and sometimes the world visits him, as was the case when Francisco Morales Maldonado, age eighty-four, visited from Puerto Rico.

Since we are all family, you can call him Frank.

From the moment I met Frank, I loved him. Yes, I realize his name is the same as my dearly departed Grandfather Frank. Perhaps that is why experiencing the positive energy surrounding Mr. Maldonado made me love him.

Billy's little strip of land provides an endless array of commonalities for those who love food and the environment. After she placed a mug with the words, I HEART PUERTO RICO, in Billy's hand, Frank's daughter, Ana Raquel, who is a frequent visitor, noticed a straw hat literally filled with just-picked figs.

"You have figs!" she exclaimed, then turned to Frank and interpreted, *"Papito, Billy tiene higos frescos!"*

"Oh, yeah," Billy said. "I noticed that the trees are full. My hat was handy, so I just loaded it up."

Billy comes from a family of music makers. Remembering that Ana had played the piano during her last visit, Billy motioned for Frank to have a seat, then said, "Ana, go over there to that pea-nanner and play us a tune."

Even though the piano hadn't been tuned in thirty years and some of the keys didn't work, Ana coaxed a joyful melody from the instrument. Soon, a customer arrived and cut Ana's recital short. While Billy tended to the customer, I led Frank and Ana on a tour.

"Where's the truck?" Ana asked while pointing at the Ford.

Shaking my head, I replied, "Don't ask."

"He still has the other truck...right?" she asked with a trace of worry etched in her tone. "I've been telling Daddy all about the truck."

Turning my back on the Ford, I said, "The real truck is out back with a load of wood on it."

As Billy continued tending the stand, I initiated Frank and Ana into the Best Friends Club. Parting the limbs of the fifty-year-old magnolia tree, I gestured for Frank to follow me into an area so special that I had only shown one other person, my daughter. Dried leaves crunched and shade replaced sunlight as we both stepped inside and stood upright beneath the bough of the majestic tree.

Smiles lined all of our faces as I explained the tree's history. "Someone told Billy it was good luck to plant a tree on your property after you moved into a new home. This was the first tree he planted."

Pointing to a makeshift cot nestled on a limb, I said, "Did you see that?"

"Oh, my goodness," Ana said while snapping a photo. "That is the perfect place to read a book."

"*Este lugar es majestuoso,*" Frank said.

Reluctantly, we left that special place for a tour of the rest of the grounds. A few short yards from the tree, we encountered the first conversation piece, one of Billy's many "rototillers." Frank smiled and wrapped his hand around the rusty metal handle. He nodded in approval and then spoke in his native tongue to Ana, who responded while I observed.

Billy appeared from behind the clothesline and said, "Now, in my day, we plowed with a mule." Using his hand to mimic the actions of placing something around his neck, he explained, "We had a yoke that we put around the mule's neck."

Frank shouted, "Yes!" He clenchted his fists together and jumped excitedly into the air.

Ana and I beamed.

"An ox," Frank said in English. "We used an ox."

Billy agreed. "Oh, yeah. Folk 'round here used an ox all the time."

The blueberry bush was loaded with yellow, immature fruit. Frank gathered a small bundle in his hands and said, "*gandules*," which, when translated, means pigeon peas. Ana, who was on the phone with her sister, shook her head and said, "*No. Son arándanos.*"

"When I was growing up," Billy interjected, "these grew in the wild. We called them huckleberries. Now-a-days folk have tame bushes and call them blueberries."

Frank's misidentification of the blueberries as pigeon peas peaked my curiosity. Pigeon peas don't grow like regular legumes; instead, they grow on a shrub-like plant that thrives in tropical climates and can live up to five years. The peas are eaten many ways: sprouted, dried, and freshly picked. Pigeon peas are high in protein. Many farmers feed spent hulls and trimmed leaves to livestock. Hulls and leaves also add nitrogen when incorporated into the soil. Finding some seeds for this multiuse plant is my top priority. Fall frosts will limit the plant's life to one growing season, but I still believe it might be worth our time to grow some. Y'all know how much Billy and I love growing new and exotic plants.

We stopped where the October beans flourished. I picked a speckled pod, placed it in Frank's weathered hands, and said, "I will dry these beans. They are like pintos."

"Ah," Frank said with a nod of his head, "pintos. I eat pintos with rice."

I responded, "Yummy. I do too."

Frank may not speak much English, and Billy and I certainly do not speak a lick of Spanish, but the more time we spent together, the tighter our friendship became. As Billy guided his guests around the woodpile, Ana noticed a vine covering a section of fence. Reaching out to touch the leaves, she asked, "Is that a passion fruit vine?"

"No. But I know what passion vines are. My mom knows how to make dolls out of the purple blooms."

In the United States, passion fruit matures to the size of a lemon. In

Puerto Rico, this fruit called *parcha* is much larger. Used in juice, *parcha* is delicious.

"We used to make candy out of the pulp," Ana said.

Overhearing our conversation, Billy said loudly, "No. Those things are dipper gourds."

I don't understand why Americans automatically speak louder when trying to communicate with someone who isn't fluent in English, but most of us are guilty of doing so. I know I am.

Lifting a leaf to display white blooms, I said, "This vine will produce dipper gourds. We have planted larger birdhouse gourds near the goats."

"Ah, gourds," Frank said. Nodding, he brought his hands together and said, "I used to make dishes and cups from gourds."

Oh, I would love to have a Frank Maldonado gourd dish. Better yet, I'd love to learn at his knee.

Arriving at the woodpile, Frank admitted, "I used wood to make a good fire when I made Pitorro, Puerto Rican moonshine."

I couldn't have asked for a more perfect segue. Leading him into the cornfield, I said, "Let me show you this.

"This is my great-grandfather's white corn," I explained. "He used it to make moonshine."

"Yes!" Frank said with another hop.

We all laughed.

"Yes. Yes," Frank said while touching the leaves. "We used shucks for toilet paper and the cobs for wipes."

Billy laid a hand on Frank's shoulder, leaned in close, and said, "Back then, we did too."

Back then, everyone did. There is no shame in our heritage. Heritage doesn't define a group of people as ignorant; it proves that they used every available resource. The lives of our people—our grandparents, parents, and other family members—make us who we are today.

Next, we showed Frank the beans entwined in the corn. He didn't understand why we had planted beans beside the corn until Billy pointed to the ground and said, "We're running out of room around here. If I'm gonna have enough space for everything, then I have to put the beans in the corn. This way, I don't have to string them up. The vines just climb up the stalks."

Again, Billy spread his arms wide and said, "This used to be the goat pasture. Now they are grazing across the street. Well, except the nannies. I keep an eye on them over here."

"How much acreage do you have?" Frank asked.

"I'm taxed at one-point-nine-tenths of an acre," Billy responded.

Ana bent toward me and whispered, "Daddy has spoken more English today than the whole time he's been in America."

Priceless. Absolutely priceless.

"Now, this isn't all my flock. Some of my goats are out in the county working."

Puzzled, Ana asked, "Mr. Billy, what do you mean?"

"I rented some of my goats to a feller over there on Etris Road. They're knocking back his weeds."

"I used to own five acres," Frank said, "but they took all of it and built a road."

"I used to have a dollhouse in the front yard," Ana added.

Billy nodded. "Here in America, they call that progress." Then, shaking his head, he added, "But it shore don't feel like it."

We all agreed. Don't we have enough roads?

"Look," Frank said while pointing toward the garden, "how pretty are your tomatoes."

And then it was a short walk from the tomato patch to the shed where Frank could finally see the Chevy.

"Look!" Frank said again while stepping toward the truck. He spoke to Ana in his native tongue while Billy almost glowed with pride.

"Papa says he has a 71 long box. He calls this a short box, because the truck bed is shorter."

All of this smiling made my face sore.

"Papa still works every day in Puerto Rico. He sells one hundred-pound propane tanks. His truck holds twelve tanks."

Glancing at this small man, I asked, "Does he have someone who helps him load the truck?"

"No," Ana said proudly. "He loads the truck himself."

Frank spoke to Ana again.

"He says he had his truck custom made. He had to add reinforcements because his truck hauls so much weight."

Billy nodded. "Oh, yeah. You got what we call helper springs."

Changing the subject, Billy said, "Now, Puerto Rico isn't a state, it's a commonwealth. But you're still a part of the United States. That makes us family…right?"

"Papa fought in Korea," Ana said. "We are born American citizens, but can't vote in the United States elections."

"Well, well," Billy said while draping an arm around Frank's shoulders. "That don't sound right, now does it?"

Frank recognized an antiquated piece of equipment Billy called a "mole board." Some also call the tool a turn plow, because it turns the dirt over and releases nutrients from deep in the soil. Billy uses this to plant and then cover seeds. Eyeing my favorite tool, Billy's hoe, Frank picked it up.

"I love that hoe," I explained. "It is a perfect fit."

"I bought that hoe in 1955," Billy said. "I knew it would last me as long as I needed one."

Then something happened that forever bonded Frank, Billy, Ana, and me as family.

Wrapping his hands around the duct-taped handle, Frank walked to the edge of the garden and hacked at the grass that was growing in Billy's tomatoes. Francisco Morales Maldonado found a piece of home on Billy's little strip of land…just as I did years ago. I was so happy I could cry.

Home. Gardens make us feel like we are home. Like the earth is ours. Frank's actions displayed respect for Billy, for the earth, for the God he loves.

Frank pointed to the sky and said, "The emblem of Puerto Rico has a green background. The green part symbolizes the farming heritage of my country. There is a Bible and a lamb on the emblem reminding us that He heals people and forgives sins. The spirit of the Puerto Rican people is symbolized in the emblem."

Next, we all made a quick stop at the sugar cane patch.

"Caña de Azúcar para hacer Guarapo," Frank said. "In Puerto Rico, we call this 'Guarapo.'"

"Momma called this 'sir-up cane,'" Billy said.

"My people call it 'sugar cane,'" I explained.

Regardless of the name, we all love it.

"In Puerto Rico, we used to grow lemon cane."

Immediately interested, I made Frank promise to try to locate some lemon cane seeds for us to grow. Wouldn't it be wonderful to have a little piece of Puerto Rico growing in Georgia? I was so happy I felt like jumping up and down like Frank.

"In the fall, we hitched a mule to the press and squeezed the juice. That's how we made molasses," Billy explained.

"We used a *trapiche,*" Frank added. "When you harvest cane, tiny hairs stick to your skin." Ripping a leaf from the stalk, he showed us the tiny, hair-like barbs. "We would chew the cane and rub it on our skin to remove the stickers. We also rubbed the stalk across our teeth to make them shiny."

Ana said, "They say sugar is bad for your teeth because of cavities, but I vividly remember chewing the sugar cane and seeing how shiny my teeth looked. Obviously it is not something you do on a daily basis."

Too soon, it was time for them to leave. Their leaving pricked my heart and caused a void, a longing to visit Frank and his homeland. How quickly I fall in love with people who exude goodness.

"Papa," I said. "I want to come see you in Puerto Rico."

I meant every word.

Frank smiled, gave me a love-filled hug, and extended an open invitation.

Maybe one day I'll make it over there. I would love to see Puerto Rico through the eyes of Francisco Morales Maldonado.

Pinto Beans

This recipe completely changed the way I prepare pinto beans. Tomatoes give the beans a rich, hearty taste that makes them the perfect comfort food on a dreary day.

Ingredients
1 cup dried beans
2 quarts water (more if necessary)
2 teaspoons salt
1 teaspoon pepper
1 small onion, chopped
2 strips of cooked bacon
1 (15 oz) can stewed tomatoes, undrained

In a saucepan, prepare beans until they are the desired consistency. Note: This process may require more than 2 quarts of water. Add salt during the cooking process.

In a separate pan, crumble cooked bacon and add chopped onions. Cook on the stove until onions become translucent. Sprinkle lightly with pepper.

Pour can of stewed tomatoes into pinto beans. Add onion mixture. Stir well. Taste to see if additional salt or pepper is necessary. Enjoy.

Red, White, and Blueberry Pie

At my house, Red, White, and Blueberry pie is a Fourth of July tradition.

Ingredients

1 graham cracker piecrust
½ cup sugar
½ cup cornstarch
1 ½ cups milk
2 eggs
1 teaspoon vanilla
1 small package strawberry gelatin
½ cup fresh blueberries
2 pints fresh strawberries

Combine ½ cup sugar and cornstarch in heavy saucepan, then whisk in the milk and eggs. Bring to a boil. Cook the custard for one minute or until thick. Remove from heat. Add vanilla. Set aside to cool.

Spoon custard into graham cracker crust. Arrange berries on top of custard.

Prepare strawberry gelatin according to package instructions. Drizzle on top of berries. Chill until set.

Note: Depending on the size of your berries, you most likely will not use all of the gelatin. Pour excess into containers and enjoy later.

25

The Hembree Grapes

In Roswell, Georgia, the Hembree Farm is hallowed ground. A sacred place. A landmark in an area that was once farmland as far as the eye could see. Around here, folk adore the Hembree Farm. Passersby love the outbuildings dotting the field, standing as a tribute to multiple generations of Hembree folk who once lived there. Truth be told, many travelers would like to see more farms and less traffic.

In 1835, the Hembree Farm, comprised of 72 acres in an area now called Hembree Road, was one of many farms in Roswell, a self-sustaining piece of property. In 1903, Pierce Teasley Hembree and his wife, Ida, built a Victorian farmhouse across from his father Robert Hembree. Their three children, Ozella, Horace, and Marie, resided with them in the home. A dirt road accommodated a scant handful of travelers each week instead of the twelve thousand commuters who utilize the paved road each day in our time.

The Hembree men were hard workers, repairing cars and working as electricians in an assembly plant. Ozella grew up to teach school and groom her garden. Time passed. The family expanded. Ozella and her granddaughter, Carmen Ford, worked together pulling weeds and snipping flowers. This is when the seed of preservation took root in young Carmen.

In those days, multiple generations lived on the same estate. As the families grew, they built homes on the property or resided in the primary home. Folk didn't live off credit cards; they lived off the land. Multiple homes dotted large pieces of property. There was plenty of room, plenty of love, and plenty of help to go around. These roots bonded families together. Today it seems like we want to get as far away from our roots as possible.

Thank God for the Hembree gals. Carmen Ford and Kelly Barkley are two of the sweetest women one could ever meet. I'm not just saying that because they allow me to pick a plethora of grapes. I say that because they love life, love the land, and are passionate about preserving history. The women live on what remains of the Hembree Farm. They work the land, tend the trees, and invest untold hours preserving structures on the property. They do good work and honor the Hembree folk of generations past. One must admire anyone who pays tribute to their heritage when they could sell out to developers.

Each fall, the Roswell Historical Society holds the annual Flea Fling on the Hembree property. The society seeks donations of goods, hosts a bake sale, offers tours of the historic building, and then prays for good weather. My role in the Flea Fling began when I spontaneously asked Carmen, whom I barely knew at the time, "How about I make some grape jelly and donate it for the fundraiser?"

"Kelly had the same idea," Carmen replied, "but she's been too busy with everything else."

That was all the encouragement I needed.

Permission to enter this sacred place comes with the acknowledgment that I should respect the memory of the residents who once lived here. Swinging a bucket in each hand, I approached a trellis that looked a lot larger up close than it did from the street. I was not alone. Hornets, bumblebees, and honeybees drunk on fermented juice flew like kamikazes, then tumbled in a contented orgy at my feet. With a silent prayer for protection, I slowly lifted a section of the vine by a single green leaf. Overripe purple hulls twisted and emitted a buzzing sound, signaling the presence of bees that at first glance appeared oblivious to my presence, but would reach out and sting me if given the chance. With a gentle shake, both the inedible fruit and its inebriated inhabitants fell to the ground.

Picking muscadines is a slow process. A flick of the knife makes collecting cluster varieties such as concords easy. With muscadines, though, their quarter-sized globes are spaced sporadically across the vine. Placing the container on the ground, I used both hands and plucked fruit from the stem. I bent and twisted, knelt and turned, working my way slowly across and then beneath the vines. When an angry bee buzzed near my head, I abandoned the bucket and bolted for safety. Giggling, I realized how foolish I must look running through the grass, arms waving. I imagined Grandmother Hembree giggling.

"That was a close one," I said while dropping another fermenting grape to the ground where it landed near a cluster of bumblebees. "There you go," I said, smiling when another bee tumbled across the ground to devour juice from my offering.

Early morning provides the best opportunity for picking. Cool, still-moist grass calms the insects and the picker. As the sun warms the grapes, hornets and other bees become increasingly active. Maneuvering among the leaves, they carefully feel each stem for a weak spot at the base of the fruit. Once a bee detects a ripe grape, it inserts a tiny, tube-like appendage called a proboscis into the fruit and begins to drink. Using jaw-like mandibles, it stretches the grape's peeling, creating an opening that grows larger until the insect can crawl inside and gorge on the heavenly juice. As the day progresses, bacteria builds up in the pierced fruit. The hotter the temperature, the more delicious the fermented juice; at least, that's what the bees think.

It is common to encounter bees so inebriated they can barely fly. On that particular day, I crossed some creatures that exhibited human characteristics akin to angry drunks. Tangling with others, they rolled on the ground, their wings vibrating a warning. Perhaps that is where the term "ill as a hornet" originated. As long as I stayed away from the drunken party, though, they left me alone.

Standing beside the vines, I listened for the Hembree folk of long ago, yearned for them to tell me a story. I believed I could hear voices of the original owners, if not for the unending sound of traffic. Even when carpool parents and school buses unloaded noisy middle school children across the street, I said a prayer of thanksgiving for this place and this opportunity.

I continued working, excited that I could use my gift to help someone else. Making jelly was an honor. I am not a check-writing kind of gal. Forgive the comparison, but I am a worker bee.

Lost in the moment, I cupped my hand and then plucked the deep purple balls. They traveled down my palm and into the bucket I held in my left hand. Inhaling the ripeness of an early season, I listened, still hoping voices from the past would visit me, but they didn't. I resisted the urge to pop a grape, or ten, into my mouth. Picking only the darkest fruit and leaving the lighter fruit for later was a challenge. My soul wanted to strip the vines clean. I did not want to waste anything, but I knew my limitations. I could not expend all my energy picking fruit. In order to make jelly, I must first make juice. If I picked all the grapes, I would not have enough stamina to process the jelly.

ꝏ

The history of the muscadine goes back to 1584, when Sir Walter Raleigh landed on the coast of North Carolina and discovered muscadines (Vitis rotundifolia). This wild grape is native to the southeastern United States. Many Appalachian residents call it a "scuppernong" or "muskey-dime." I imagine Sir Walter Raleigh examining the fruit and smelling it. I know his mouth watered when he inhaled the ripe fragrance. He smiled when he tasted it. Of that I am certain.

Today, wild muscadines wind through the Carolinas and most of the South, providing food for wildlife and foragers such as myself. A thick, almost leathery, skin protects the seed-filled, pulpy center that, when crushed, renders a rich, sweet juice. Muscadine fruit doesn't grow like bunch or cluster grapes. The individual grape is larger, and the yield from vines is small when compared to other varieties. According to the Georgia Cooperative Extension service, the state of Georgia is the largest producer of muscadine grapes. Approximately 1,400 acres yield a crop used for jam, sauces, and wine. The grape contains resveratrol, the component in red wines that lowers cholesterol and the risk of coronary heart disease. Recently, the *American Journal of Enology and Viticulture* (vol. 47, pp. 57–61) reported that two ounces of unfiltered muscadine juice contains the same amount of resveratrol as four ounces of red wine.

Let us lift our glasses and say, "Cheers."

Those living in the Southeastern United States can easily grow grapes. Cultivation is a simple process that yields long-term rewards. Homeowners can incorporate vines into their landscape, using them to cover a fence or trellis. Plant the vines in well-drained soil. Vines need full sun and will die if exposed to standing water, even for a short amount of time. Erecting a trellis prior to planting is encouraged. After constructing a trellis, place the root ball in the center at the base of the support post, and then train the vine to grow straight (not wrapping around the post). Prune shoots the first two years. Once the tendrils grow to the top of the trellis, encourage them to grow laterally. This may mean tying the vines to the wire.

Many Cooperative Extension offices sell varieties of grape vines and offer literature about everything from growth to food preservation. The University of Arkansas distributes informative literature titled *Muscadine Grape Production in the Home Garden* by Dr. Keith Striegler. Those considering expanding their gardens to include this delicious fruit should locate the online article and read Dr. Striegler's suggestions (http://www.uaex.edu/publications/pdf/FSA-6108.pdf).

I strongly encourage homeowners to plant muscadines. They ripen in early fall, providing abundant and beneficial fruit. If I had room, I would plant several rows of grapes on my property. Unfortunately, I don't have even an inch of soil that receives full sun. Yet I remain a loyal fan of the muscadine. Each delicious taste nourishes the body and replenishes the soul.

Grape Juice

This recipe is for those who wish to preserve the juice.

Ingredients
3 ½ pounds of grapes, washed with stems removed
Pure cane sugar to your taste
1 ½ cup water
Clean jars

Pour grapes into a large pot, and add approximately 1 ½ cups of water, or enough water to cover fruit. Bring grapes to a boil, and process for 5 minutes. As the liquid begins to boil, add a half-cup of sugar. Use a pastry cutter, potato masher, or slotted metal spool to crush the grapes, and then remove a spoonful of grape juice and taste. Add more sugar if desired.

Drain juice through a colander. Using the pastry cutter, press grapes to release residual juice. Note: Those wishing to capture antioxidant-rich resveratrol pulp should proceed to the bottling process. Those desiring clear juice with minimal pulp should strain the liquid several times using cheesecloth and then proceed to bottling process.

Bottling Process

Pour mixture into clean glass jars, and secure with lid and rings. Process the juice using the water bath method according to container size and altitude. The National Center for Home Food Preservation recommends the following:

Container Size	Altitude	Process Time
Pints–Quarts	0–1,000 ft.	5 minutes
Half-Gallon	0–1,000 ft.	10 minutes
Pints–Quarts	1–6,000 ft.	10 minutes
Half-Gallon	1–6,000 ft.	15 minutes

Refrigerate any jars that do not seal, and consume juice immediately.

Grape Jelly

In order to make jelly, you must first juice the grapes.

Ingredients
5 cups grape juice (using Grape Juice recipe)
7 cups pure cane (not granulated) sugar
1 teaspoon lemon juice
1 box powdered Sure-Jell or other pectin product
Metal spoon (to test jelly)
1 box of 6-ounce jelly jars with lids

Note: These directions are for traditional Sure-Jell, powdered pectin, not liquid. Other brands of pectin may feature different directions. Refer to the package insert prior to processing.

Place metal spoon in the freezer. You will use it later to taste the jelly.

Pour grape juice in large pot and set temperature to medium high. To prevent jelly from turning dark, add one teaspoon of lemon juice and one container of pectin. Stir continuously until juice comes to a full boil. This takes several minutes.

When the mixture reaches a boil, add 7 cups sugar and stir well. Bring mixture to a boil. Boil for one minute.

During this process, a bubbly foam floats to the top. Skim this off and discard. You will feel the jelly thicken slightly as you stir. Do not expect the jelly to harden until after you have placed it in the jars.

Remove the metal spoon from the freezer and use it to test a small amount of hot jelly. A cold spoon rapidly cools the jelly and allows you to test the consistency. Jelly should slightly adhere to the spoon.

Pour molten jelly into jars, seal, and process 5 minutes using the water bath method.

26

Reflecting on Farming and Friends

There was a time when gardening embarrassed me. City slickers don't grow vegetables; they buy them. Just like they pay people to manicure their lawns, design a garden fit for *Southern Living*, clean their homes, and tend their children. That's why they live in the city, away from farms. They work hard in their starched, white-collar jobs instead of staining their clothes with sweat and grime. As the saying goes, I've been there, done that. I will probably never make more money writing books than I did during the time I call "my other life," the period when I had a real job and a retirement account. Some days, I mourn that life, miss the satisfaction of a job well done, miss my colleagues. Being an author is a lonely profession, painfully so for outgoing people like myself. Being a farmer is a lonely job too. I think that's why God put Farmer Billy in my path. Billy has introduced me to a lot of real friends—those I can call in the wee hours of the morning, who love me and my family, who give hugs, share recipes, and bow their heads in prayer for me.

Writing the first book, *In the Garden with Billy*, allowed me to experience the kindness of others. Touched by the story, strangers reached out. They wrote letters telling me that I had reminded them of their grandfather or their childhood. Inspired, they wanted to help Billy or their neighbors. These strangers became friends. As Mr. Coleman once said, "A body can never have too many good friends."

When I decided to focus on my writing career, my husband constructed deck planters so I could incorporate a "bloom where planted" methodology. I do a lot of thinking while touching the soil. I strike the earth hard when I'm frustrated, afraid, homesick. I crack the ground open, argue with the clay, dare it to defy me when I plant a bulb, seed, or shrub. I have a lot of Grandma Wonderful's DNA running through my veins. She gives her flowers two growing seasons and then, with a tug and a rip, you're outta here. Let's give something else a chance.

My first year in Atlanta, I sat on my sofa and watched wave petunias bloom in the deck planters outside the window. I wanted an outdoor living area like the Joneses. Today, plants trellised using bamboo teepees promise a tomato sandwich. In the fall, garlic and lettuce grows dog-hair thick. Two steps from my kitchen door, fresh vegetables wait for me. I have filled the raised beds my husband worked so hard to perfect with vegetables. Conjure up an image of Granny Clampett, and you have a good idea of what my outside living area looks like. Most visitors see what the deck could be, what my husband hoped it would be, a canvas upon which to cultivate a rainbow of blooms.

Sometimes I wish that were the case, wished that perennials bloomed with little maintenance; that I had created an area butterflies and birds could call home. But there are other times, moments when I catch my husband looking at me, peeking through the thirty-year-old windowpane as I work in my tiny garden. I cherish those moments. I lock the look on his face deep inside my heart. There, I saw it, a crinkle in the corner of his eyes as a smile forms. We are proud that love lives in our atypical Atlanta home. Manicured lawns and sprinkler systems would stifle us, force us to conform and acknowledge that we are just like the rest of the city slickers.

We are not.

We could strive for perfectly pruned crepe myrtles and have someone clip the grass exactly two inches high while I get a mani-pedi. We could pay someone to build a patio, install a pool, and keep our outdoor living space even with the Joneses. Or we could let the seeds fall where they may.

I could plant wave petunias, or I could plant love. Both require a lot of tending. I think I made the right choice.

ᴓ

Summer breathes hot across the South, wilting crops and punishing livestock. Billy Albertson struggles during this demanding time. The lawn grows too fast. Laundry piles up. Dirty dishes clutter the kitchen. Then there are the vegetables, which ripen faster than he can pick.

During the months of June, July, and August, Billy Albertson needs help. After Billy's wife passed away, his daughters hoped a Boy Scout Troop would adopt their father. Instead, they got me, a gal who is five feet tall and, at best, can only work three days a week. Others help Farmer Billy when their schedules allow. Truly, Billy and his daughters appreciate all volunteers.

For Billy, a typical day begins before sunup. He doesn't sleep well at night—another side effect of hormone treatments for prostate cancer. He converts sleeplessness into bean-stringing opportunities. He does a load of laundry and cooks up a week's worth of breakfast that he stores in the refrigerator.

Commuter traffic packs Hardscrabble Road. I signal, pull into Billy's drive, and park beside his brick home and away from the circular driveway. I always leave the driveway unobstructed. Blocking the road is one of only a few things that upsets Billy Albertson.

"Folk around here don't know how to back their vehicles up," Billy explained shortly after our introduction. "That's why I built the circle out there." I chuckle at the board marked TOLL ROAD. Janet made the sign for his eightieth birthday. Billy had so often joked, "I should charge folk fifty cents for pulling in here. If the state gets away with charging folk to use their road, I think I could too."

By 8:00, the smell of chicory coffee hangs so strong that I can almost scoop a cup from the air. Billy has completed what he calls his domestic duties and critter chores. Animals munch contently on grain, and this simple man ponders which section of the garden to enter first. Most days, Billy's animals eat before he does. I'm ready to go, chomping at the bit to hit the field, pick, sack, and be done for the day. Billy is in no hurry. With callused hands wrapped around a coffee mug, he slurps the java, chews on a sausage biscuit, and then slips his feet into a pair of well-worn boots that he calls "brogans."

Observant visitors notice the staples and tiny nails hammered into the rubber soles. Some might even recognize a hint of duct tape holding the shoes together. Duct tape and WD-40 are necessities on the Albertson farm.

After rounding up all available empty buckets and piling them in a wheelbarrow, we enter the garden ready to pick. If Billy knows that I am coming, and I am usually there every Monday, he heads toward the okra patch, leaving me the tomatoes. I pull my hair into a ponytail and tighten the Velcro on my sun visor. When I first began helping Billy, he placed an empty bucket in each of my hands, pointed me toward the field, then followed behind pushing the wheelbarrow. I picked one side of the plant, he the other. Today, the harvesting process has advanced with the invention of the "tomato toter."

"I like to think this will help us work a little smarter," Billy explains as I examine the apparatus. "Go ahead," Billy says with a smile, "give 'er a try."

Comprised of sturdy wire screening nailed to pressure-treated lumber, the toter has a rectangular shape that covers the frame of a recycled wheelbarrow frame with a missing tray. I assume that the tray finally succumbed to rust, although I bet my bottom dollar that, if I looked hard enough, I would find it lying somewhere on the property. Firmly secured with bungee cords wrapped around the wooden frame, the toter isn't exactly aerodynamic. It is, however, functional. Watching someone steer the toter is entertaining, which is why Billy stands back and gives me plenty of room.

Tightening my grip around mismatched handles, I push toward the narrow pathway.

One handle is "standard issue," factory made, and the other is a wooden two-by-two. The lack of consistency in weight and texture throws me off balance.

After maneuvering between the out buildings and chicken lot, I dodge a variety of empty containers, discarded tools, and tractor parts. This contraption has a tendency to tip forward, even when empty. Complicating the journey is a winding path with a deceptive stretch of overgrown grass that hides a drainage ditch. Billy's plot is lower than that of his developed neighbors. Water stands between the rows after heavy rains. During the winter, he routes runoff through a trench in the field. Sinking a wheel into the trench alters the balance of the toter. Fortunately, the toter is empty when I bust the ditch wide open. Rocking the device back, then forward, and finally giving it a hard push, I emerge from the mire and make a mental note to lay a board across the hole so I can cross next time with minimal difficulty.

Billy still watches me from a distance. It may appear that he is lost in his work, but trust me, that man has one eye on the okra and the other on his worker. I pluck tomatoes and place them directly on the screen, piling vegetables higher and higher until I can barely steer. Billy appears, takes command of the toter, and wheels it to an outdoor rinsing sink. He cleans the tomatoes and then pushes the device to the carport, parking it beside the certified scales.

"You dry. I'll sack," Billy says while handing me a towel.

By now it's 10:30. A five-gallon bucket is full of okra, promised to a customer before the pods have developed on the stalk.

"This is in high demand," Billy tells me while placing the bucket in my hand. "Set this in the house. I've called someone to come pick it up. We've got to get it out of sight before customers show up."

Working without words, we slip into our roles. I struggle to keep up with Billy and opt to dry the tomatoes by collecting several into a towel. Drying each red globe takes too much time.

I don't need instruction during the sacking process. Each bag contains one large tomato, several medium-sized ones, and two or three small ones that bring the weight up to two and a half pounds. Tomatoes aren't sold individually, or folk would take all the large ones, leaving an overabundance of smaller fruit. Knowing this, I arrange according to size so that Billy doesn't have to hunt and peck for what he needs.

Even though the familiar red tomato-shaped sign still rests against the side of the house, Billy's regular customers arrive. I swear that some people keep the road hot driving back and forth, hoping for a glimpse of him.

Interruptions are common. Thus the need for a self-serve payment method, a dented Charles Chip Bucket marked HONER SYSTEM.

"Just toss your money into that ATM bucket," Billy says to a customer. No time for chitchat; there are tomatoes to sack. "Let me know if you don't find what you need."

Now that I am familiar with the flow of customers, I make myself scarce after Billy places the sign at the end of the road. He needs to visit, socialize with his friends. My escape allows me a moment of solace. Turning a bucket upside down, I hide in the bean patch. My mind wanders. Green tendrils creep up cornstalks. Time slows. Bees pollinate delicate white and purple blooms. Zipper peas grow fat in the sun. I am thankful for this moment, this feeling, this peace. I wish everyone could experience such joy. It makes me sad when others can't experience the garden as I do.

By 11:45, it is time to start thinking about the meal Billy calls "dinner." Anticipating Billy's hunger, I stop working and assemble a twelve o'clock sandwich. Otherwise, we complete the planting, picking, and pilfering in the order he deems necessary.

"I don't make plans," Billy always says when I ask what's on tap for the day. "Just head out and do what needs doing."

Dinnertime is an exception. Most days, Billy wants to eat at noon. Preparing for an influx of working folk who drop by the stand during their lunch hour, Billy stocks the table with tomatoes. Satisfied that the inventory is adequate, he steps inside to wash up right about the time I place a slice of fried bologna on the bread.

"Ho there!" he shouts from the kitchen sink after seeing a customer through the window. "You eat yet?"

I peer out the window and grab another tomato to slice. Regardless of the answer, his next statement is always the same. "Step on in here and we'll make you up a sammich. We've got plenty."

Wise customers accept this offer.

By one o'clock, I am exhausted. After we have finished eating, he says, "Let's sit down in the living room and take a breather. We've been going at it all day."

This is my cue not to join him but to manufacture an excuse to let him rest. What Billy really needs is a nap, and I do too. This is why I leave after dinner. Billy eases into a lawn chair positioned beneath the pecan tree. He will rest a spell and then, later in the cool of the evening , he will work a spell. Tomorrow, Lord willing, he will arise from the bed and be about his business again.

This is Billy's cycle of life. Work hard, rest hard, love hard. Spend time with friends and neighbors.

After working six days straight, Sunday is his day of rest. Visitors seeking uninterrupted Billy time arrive, bringing food and friendship.

For those considering a visit to Billy's, please remember that he does not exchange money on the Lord's Day. Nor does he pick produce or allow you to pick anything from the garden on Sunday.

But if you want to sit and visit a spell, you're more than welcome to pull up a lawn chair beneath the shade of the pecan tree.

Veggie Love

Reading Group Questions and Topics for Discussion

1. The author writes about saving vegetable seeds and planting them each spring in honor of loved ones. What are your most cherished memories of your loved ones? How do you honor their memory?

2. In the chapter titled "Cotton Is King," are you surprised that the author and Billy choose to plant cotton on the farm? Have you ever bent the rules? If so, what was the outcome?

3. The author shares some of her favorite recipes. What are your favorite food memories?

4. In the chapter titled "Jill," Billy's daughters confront the woman who took the goat named Jill without paying. Would you have done the same? Or would you have been like Billy, who decided to wait and see if the woman would do the right thing?

5. Billy sometimes puts his customers' need for produce ahead of his personal health. How would you suggest Billy maintain a balance that keeps his customers satisfied and allows him to rest when necessary?

6. Farmer Billy is surrounded by neighbors who accept his rural lifestyle and sometimes help on the farm. Would you like to live next door to him?

7. Billy Albertson welcomes strangers from across the world onto his property. Has his eagerness to make new friends changed the way you look at your neighbor? If so, why?

8. When the author tries to control kudzu on her property, her efforts take an unexpected turn. Have you ever tried to resolve a situation only to watch helplessly as your effort spirals out of control? How did you handle that situation?

9. Hidden on the Albertson property is a tree house where a group called the "Best Friends Club" meets. Growing up, did you have a tree house or a secret club? Discuss.

10. The author is passionate about supporting local businesses and fledgling entrepreneurs. Do you support small businesses in your area, such as independent booksellers? If not, would you consider giving them your business? Why do you think it might be important to do so?

11. With the move by some consumers to buy locally grown produce, has your opinion of where your food comes from changed? Would you consider growing your own vegetables?

12. The author mentions eating fried egg sandwiches that her Grandma Wonderful makes. Share with the group some ways that food made you feel special when you were growing up.

13. Were you surprised when the author confronted the intoxicated customer who had picked corn without permission? How would you have handled the situation?

14. After reading *In the Garden with Billy*, Neisha Handley and other readers reached out to strangers like Billy Albertson. Some began visiting the elderly in their neighborhood. Does this book make you want to step outside of your comfort zone? Why or why not? What are some ways that you can do this?

Recipe Index